Institutional Governance and Regulation of Water Services

The art of progress is to preserve order amid change and to preserve change amid order

Alfred North Whitehead
British Philosopher
1861–1947

Institutional Governance and Regulation of Water Services

The Essential Elements

Second Editon

Michael Rouse

IWA Publishing
London • New York

Published by **IWA Publishing**
Alliance House
12 Caxton Street
London SW1H 0QS, UK
Telephone: +44 (0)20 7654 5500
Fax: +44 (0)20 7654 5555
Email: publications@iwap.co.uk
Web: www.iwapublishing.com

First published 2013

British Library Cataloguing in Publication Data
A CIP catalogue record for this book is available from the British Library

ISBN 9781780404509 (Paperback)
ISBN 9781780404516 (eBook)

Contents

About the Author

Michael Rouse is a Distinguished Research Associate at the University of Oxford and manages the Institutional Governance and Regulation module of the University's MSc Course on Water Science, Policy and Management. He was formerly Head of the Drinking Water Inspectorate in London and has extensive knowledge and experience of water governance and regulation, including all aspects of audit and enforcement, and of the governance issues related to both public sector management and privatisation. He has wide knowledge of water technical and operational matters, based on his applied research and development background at the Water Research Centre, where he spent 9 years as Managing Director. Michael has a good understanding of international water matters and works independently as an advisor to governments on policy and regulation. He is a Past President of the International Water Association, a member of the IWA Council of Distinguished Water Professionals and a Past President of the UK Institute of Water. Michael is an Honorary Member of the American Water Works Association, a Foreign Associate of the US National Academy of Engineering and an Honorary Member of the American Academy of Environmental Engineers. He is a visiting professor at Tsinghua University in Beijing and at the Shanghai Academy of Social Science. In 2000 he was awarded the CBE (Commander of the British Empire) for his professional services.

Acknowledgements

I am sorry that I cannot list the names of all who have helped to shape my views over the years, but I am grateful for their collective wisdom and experience. Those who work in the world of water are dedicated people, passionate about the service they provide, and willing to share their experience to the benefit of others. I thank them all, and am grateful for the opportunity and good luck to be part of that international water community. I hope that at least some of them will feel that I have done them justice in trying to assemble the collective wisdom on governance, policies and regulation.

In my international work, and during my time as President of the International Water Association (IWA), I have been privileged to meet many excellent people involved in water. They come from different types of organisations: government, non-government organisations, regulation, local government, public and private utilities, trade associations, international associations and academia. What these many contacts have taught me is the importance of looking at situations from a new perspective without preconceptions. Each type of organisation approaches issues from a different viewpoint, and each contributes to the establishment of effective policies.

It is inevitable that the greatest influence has come from my colleagues, firstly at the Water Research Centre (WRc), then at the Drinking Water Inspectorate (DWI), and in recent years, from the Oxford Water Research Network at the University of Oxford. I am not listing names for fear of omitting someone, but I thank them all. On regulation, I include my former colleagues in Government, at Ofwat and in the Environment Agency. These pioneers have made a highly significant contribution to the development of water regulation.

IWA organises workshops in a number of countries during which water industry leaders discuss issues faced in achieving sustainable water services. The presentations and discussions include governance and regulation. During the period I was involved, I learned a great deal from both the team of international presenters and from the local leaders. I am particularly grateful to John Langford, Paul Reiter and Mark Pascoe.

I give special thanks to Rachael McDonnell, the main architect of the University of Oxford MSc Course on Water Science, Policy and Management, for her vision and courage in inviting me to be a member of the teaching team back in 2004, and for providing me with stimulus and encouragement. My preparation for teaching the module on institutional governance and regulation involved rigour and discipline in research. The resulting structure provided the basis for the first edition of this book. The students in years 2004–5 and 2005–6 of the Oxford MSc Course provoked me into writing the first edition. I am grateful to those students, and the following alumni, for their stimuli and for their probing questions, which cause me to review and refresh my thinking. I pay tribute to my late wife Penny who made a vital contribution to my work over the years and in the preparation of that first edition.

Many colleagues have contributed to the new material, and in the preparation of this second edition. I mention particularly Wisdom Akatu, Sanford Berg, Alan Booker, Samir Ghith, Helene Horth, Anand Jalakam, Piet Jonkers, Maria José Pallares Paredes, Mark Pascoe, Nick Pilgrim, Paul Reiter and Han Wei. I thank them all. Contributions have been made by the MSc students in their dissertation projects. Their work has added to the knowledge base and reinforced many of the key messages in the book. I am most grateful to Rob Hope, Director of the MSc Course over the period 2006–13, for his strong support and for providing me with a valuable research base.

Michael Rouse
May 2013

Foreword

The complex and often politicized nexus between – (1) the governance of water services including its regulation, 2) the organization and management dimensions of efficient delivery of these services and 3) the roles of the public and private sector as service providers – is a complex subject. This is an unfortunate truth – because getting the nexus right is critical to the efficient and effective supply of water service, in developed or developing countries, and in big and small utilities alike – and getting it right is not easy.

More often than not, the complexity of the nexus results in books and articles that resort to reductionist arguments about both the problems and the remedies to this challenging nexus. For example, some argue that privatization will cure all, others argue that regulation is the salvation while others argue that efficiency, portrayed as no price increases ever, is the ultimate target.

What is so important and unusual about this book, now in its second edition, is that it tackles the whole nexus of these intertwined issues in a clear, balanced and comprehensive manner. Michael has drawn on his extensive working experience on these issues to produce cases that illustrate both the principles and the solutions associated with good governance and regulation leading to efficient and effective water service provision.

Personal experience in both running multiple utilities in Seattle and then later, having interacted with several hundred utilities worldwide through my position as Executive Director of IWA, has convinced me that the subject of this book is a journey – a journey that continues over some considerable time. Great progress can be eroded through political forces – witness examples from Australia related to the drought. At the same time, rapid progress occurs in unlikely circumstances using Rumania as an example. In the aggregate, we are moving in the right direction.

If we are on a journey, which I believe is the reality of our situation, continuing to learn and advance on this path to efficient and effective water services is vital – vital to individual well-being and vital to the economic health and well-being of countries and regions. And if we are to address the already huge gaps in coverage of safe and reliable water in the vicinity of home, while meeting the water needs of over a million new residents to planet earth every week, we need to dramatically pick up the pace on this journey.

In this context, we all need to thank Michael Rouse for his great contributions to this field, and for writing this very useful book, not just once, but twice. And we as water professionals need to commit to using the lessons and principles provided through this book in a much more vigorous manner, both within IWA and in the industry at large.

Paul Reiter
Strategic Counsel to IWA

Preface

What's it all about!

Water is essential for life, and globally the provision of water services is the most important of all functions, yet so often the policies and governance structures necessary for its effective delivery are not in place.

Effective institutional governance and regulation of water services are essential for the establishment of sustainable systems, yet at the time of writing the first edition ahead of the year 2007, I was not aware of any books which covered the subject comprehensively. In developing the module on institutional governance and regulation for the University of Oxford MSc Course on Water Science, Policy and Management in 2004, there was little published material at that time to assist me, and I had almost to start from scratch. I am pleased that since then the subject has received much greater attention.

I was fortunate in having many contacts within the International Water Association, with whom I discussed the regulatory issues, and who provided sources of valuable case study information. My nine years experience as one of the Regulators, working within the regulatory structure in England and Wales, provided me with a sound appreciation of some of the fundamental principles, and my own later studies identified a number of key factors for successful regulatory systems. My more recent work, on regulatory matters in other parts of the world, together with my work with colleagues at the University of Oxford, and students' dissertation projects, has reinforced my understanding and beliefs in what are likely to constitute successful structures and regulation, to achieve sustainable water services. The examples in the book are based largely on water supply, but I believe that the principles apply equally to wastewater services.

I have included a new chapter (Chapter 1 Part2) on water and sanitation as a human right. This is influencing thinking on what should replace the millennium development goals for water and sanitation after 2015. The UN Declaration has the potential to provide the stimulus for substantial progress towards universal coverage. The chapters in this book are relevant to specific aspects of that declaration.

At the time of writing the first edition, UNDP published its Human Development Report 2006 (*Beyond Scarcity – Power, Poverty and the Global Water Crisis*). This provided an excellent analysis of deprivation of access to water and sanitation for billions of people. The report gave a great deal of emphasis on the need for good political leadership, and for good governance and regulation. In this second edition, I retain the appendix on the report, mainly because the recommendations are as valid today as they were in 2006, but also as an example of work, which although acclaimed at the time, has largely been forgotten. The "wheel" continues to be re-invented. This second edition is about good governance and regulation, and throughout the book I have attempted to cover both similarities and differences in requirements between developed and developing countries.

I don't pretend that this book is an exhaustive analysis of the subject. It is based on common-sense principles and what works well in practice. It is clear to me that effective and sustainable water services are wholly dependent upon effective institutional structures, regulation and good management. Good understanding of the science of water, wastewater and the water environment is very important, but that science cannot be applied successfully unless effective institutional and management structures are in place. Establishing those vital structures is more difficult, because it does not involve physics, chemistry or biology, but people. Some of the people involved will have their own specific well-meaning objectives, which may be in conflict with others. Achieving common objectives requires strong and often courageous leadership by those in a position to establish the necessary institutional governance and regulation arrangements. This book aims to assist those leaders, their advisors and their staff. I hope also that tomorrow's leaders, today's students, will find that the book assists them in their studies, and that it will play a part in stimulating them to pursue careers in water and sanitation.

Chapter 1

Part 1 – Important policy issues

Some bricks in the wall

This book is about the good governance necessary to achieve effective, efficient and sustainable water services. So what constitutes good governance? Good governance is taken to have various meanings in definition and scope. In 1997, the United Nations Development Programme (UNDP) defined good governance as "The exercise of economic, political and administrative authority to manage a country's affairs at all levels. It comprises the mechanisms, the processes and institutions, through which citizens and groups articulate their legal rights, meet their obligations and mediate their differences" [1]. To me, in the context of sustainable water services, it means that at least the following should be present:

- Government structures in which sound integrated policies can be developed;
- The professional capability within governments and civil services to develop sound policies;
- Separation of policy, regulation and delivery;
- Regulatory bodies that are independent and that have the power to implement policies, laws and rules in a transparent way;
- Effective monitoring and enforcement of regulations;
- Open and transparent information and reporting as the most effective means of combating corruption;
- Effective public participation as part of transparency, but also to provide for buy-in and commitment;
- Utility operators who, having been given clear objectives, targets and resources, are free from political interference, so that good operational management can concentrate on providing efficient services.

Without good governance, very little can be achieved, and it is not possible to utilise good science and technology effectively. Indeed, poor governance is worse than no governance at all, as wrong decisions can not only prevent the attainment of the desired water supply and sanitation goals but can result in a reduction in the delivery of water services. It is therefore appropriate for this chapter to deal with government, structures and policies, to give an overview of the challenges in achieving sustainable water services. I make no excuse for covering critical items in this chapter, and expanding on them in later chapters, as repetition provides for emphasis.

GOVERNMENT STRUCTURES AND RESPONSIBILITIES

Overall responsibility for policy generally rests with central governments although in some countries, such as in Germany, it is delegated to a regional (Länder) level. Important policy issues range from decisions on governmental departmental structures, to consideration of more detailed aspects such as who should be responsible for drinking water quality in public buildings.

Many government functions have an interest in water. Water relates to health, to the environment, to agriculture, to regional development, to housing and to finance among others. Because of this, there has been a tendency for the responsibilities to be spread across government departments. Drinking water often comes under the Ministry of Health, with wastewater being the responsibility of the department responsible for the environment, as in Sweden. In some countries of the world, which are highly dependent on irrigation, water is attached to the Ministry of Agriculture, as in Saudi Arabia, Montenegro and Namibia. Water resource development can be separated from responsibility for water supply and sanitation services, as in parts of China.

In the past, perhaps, this didn't matter very much but with the increasing demand for water, and the need to include wastewater re-use as a part of water resources, there is a need for an integrated approach to policy and planning. It is highly desirable that just one government department is involved as in the UK but, if not, there needs to be a designated lead department, or some form of coordinating committee with sufficient status to overrule territorial obstacles to progress.

It is estimated [2] that the world's population will increase from 7 billion to 9.3 billion by 2050. Because of increasing urbanisation most of the additional 2 billion plus will impact urban areas. The situation will be compounded by the population living in smaller units [3]. This will challenge all aspects of urban development with water and sanitation provisions becoming an integrated part of sustainable cities and towns. Consideration has to be given to how city planning is reconciled with administrative structures based on regional and hydrological boundaries.

The water and sanitation needs of rural areas will not diminish, as access levels trail those of urban areas, but the problems are different.

There is then the question of regulation. Who decides on water charges? How should the environmental "costs" of water be reflected in water prices? Who monitors against the required deliverables? Who has responsibility for enforcement of standards? Who will issue reports on drinking water quality and environmental standards? Can government, whether national, regional, or local, be responsible for policy, and also be responsible for open and transparent monitoring, enforcement and reporting of the impact of those policies?

SEPARATION OF POLICY, REGULATION AND DELIVERY

The traditional structure for water services is the municipal model in which central government, in consultation with local government, determines overall policies, and local government is responsible for detailed aspects of policy, for example policy on cost recovery, and for delivery and monitoring. Often, the municipal utility monitors its own performance. Water charges are set by local government, with little or no transparency, where in many cases the water charges are included in local taxation. Consumers don't know how much the service costs or whether or not it is an efficient service. In some cases, water charges are separate but consumers don't know whether the income is being used wholly for water services, or whether it is used to subsidise other local services such as transport.

Government's role is policy making, and those involved in government are generally most suited to a policy function. Elected local government officials are interested in politics and policy, and local authority staff have been trained as administrators, not as operational managers. There are exceptions but governments, either national or local, are not generally good at managing operations, and with policy responsibilities having to take priority, the direction and energy of the officials is inevitably taken up in local political matters. The loser is efficiency of utility operations. The solution is to separate the delivery functions, either by creating operational bodies within the local authority, such as in Seattle in the United States (see Chapter 7), or to form city-owned companies operating independently, as in the Netherlands, to contract out management to the private sector as is common in France, or to completely separate the functions through the formation of equity-based asset-owning private water companies as in England and Wales. The separation has to be complete and transparent with government officials unable to interfere in operational matters, for example in staffing levels. Each form of separation should be accompanied by transparency of budget and accounts, and effective monitoring of the required deliverables.

In each form of separation in which the municipality retains ownership, it can be regarded as the parent, and the operating unit as its child. Parents can be very indulgent of their children, and reluctant to publicise any misdemeanours. Therefore, in order to achieve effective scrutiny of the operations, it is necessary to introduce some independent body to act as a regulator and auditor of published results. This doesn't mean that there has to be a body called a regulator, but it does mean that there should be a body which is independent of the local authority and the operator, and which has the authority to take enforcement action to require the operator to rectify deficiencies in performance. This is a big step for a local authority to take, as it means handing over control of information, and generally it can only be achieved by national legislation or dictate.

The benefits in performance will be seen in Chapters 7 and 8 where examples of separation of policy, regulation and delivery are discussed. However, in order to attract high quality management to publicly-owned utility companies they need to be of sufficient size, both to offer attractive salaries and to have the required scientific support functions.

THE QUESTION OF WATER UTILITY SIZE

On the question of effective water service utility size there are conflicting influencing factors. Historically, in the developed world, water service utilities were set up by communities to meet their own needs. Some of these communities grew into large towns and cities, and the capability of the utilities grew accordingly, but others remained small and rural. In parts of the United States and in Canada there are a large number of small units supplying less than 5000 people. These benefit from local civic interest and pride, but often the increasing demands to meet higher standards, and replace ageing infrastructure, are beyond the financial and management capability of these small utilities. There is an understandable reluctance by the communities to give up direct control of their water services, but the small utilities are unlikely to be able to operate in the future without financial and professional support subsidies. In Scotland in the UK, the approach has been to form one large public utility operator, Scottish Water, covering water and wastewater operations for the whole of Scotland. This provides all consumers in Scotland, whether urban or rural, with the same benefits of high quality professional staff and the finance necessary to meet drinking water quality and wastewater standards, and to rehabilitate the ageing infrastructure. The consumers in Scotland may point to increased water charges, but there has to be recognition of the need for realistic charges for sustainable water services. This is critical for success and is covered later in this chapter and in Chapter 3.

In the developing world there is a different situation, with emphasis in many areas on achieving access to basic services. The Millennium Development Goals (MDGs), whatever their measurement and monitoring limitations, provided stimulus for action. The Joint Monitoring Programme Report 2012 [4] reported success on meeting the water goal but not on sanitation. On 28 July 2010 the United Nations General Assembly adopted a resolution "recognizing the right to safe and clean drinking water and sanitation as a human right that is essential for the full enjoyment of life and all human rights" [5]. This resolution will give greater impetus to the challenge of progressively achieving "universal" access to water and sanitation. However, there could be a number of pitfalls in the implementation of the resolution. These must be addressed by governments in determining policy to prevent the loss of this opportunity for progress. This important subject is discussed in the second part of this chapter. In addition to government action it is necessary to harness the energy of those people needing the service at the local community level. The highly successful NGO, WaterAid, is pursuing its Citizens' Action Project [6]. There has been success in doing this in small rural communities, and the challenge is to achieve the necessary community leadership and commitment for communities of 100,000 and more in towns and peri-urban areas. Such an approach encourages people to look for local solutions regarding water resources, incorporating rainwater harvesting, groundwater recharge and re-use of wastewater. The size of these operations is highly community-specific, but there is still a need for good professional management to design and operate the systems. In parts of Brazil this has been solved by forming co-operatives, where a number of communities have come together to form larger viable management units under the supervision of an elected committee. [7] This doesn't mean that the individual supply units need to be joined together hydraulically. The benefits come from being able to employ high quality management and from having the ability to raise capital for system improvements. Levels of public participation and examples of successful involvement are given in Chapter 5.

CENTRALISATION OR DECENTRALISATION?

Size is just one factor in decisions having to be taken by countries on the degree of centralisation or decentralisation. It can be argued that all decisions, whether they are on water resources or quality standards or financing and cost recovery, can be taken at a community level. The converse argument is that water resource planning and decisions on environmental standards need to be taken on the scale of river basins. In 1974, The UK Government decided that all public water supply and wastewater operations should be combined into just 10 Regional Water Authorities (RWAs). The geographical areas of the RWAs were watershed boundaries to give a water catchment approach to water management. This change was very unpopular within local authorities whose utilities were taken away from them, and proved to be very controversial with the general public as, in many areas, they received separate water bills without a commensurate reduction in local taxation. The bold decision was lauded by the world's experts at that time as the ultimate in water management [8]. The RWAs were overseen by central government appointed Authorities, initially made up largely of local authority representatives, but later smaller more commercially-orientated boards. In one sense this could be considered as a centralised approach with the integration of a large number of small operators. However, as the RWAs were required to become financially independent, state grants ended, so it could be argued that there was decentralised delivery albeit through large utility units.

Within the European Union (EU) centralised policy making has been taken one step further with Member States deciding collectively on standards. The Water

Framework Directive [9] sets mandatory goals on environmental water quality, and requires the formation of Responsible Authorities to manage river basin systems.

In India, although the National Water Policy of 2002 [10] sets down a broad policy framework, water is a state responsibility and the states are required to develop their own policies based on that framework. The framework includes reference to river basin management and implies that large water resource developments will be required. This has resulted in vigorous public debates on the direction required in particular states. The policy document refers to the use of all available sources of water and, I quote, "Non-conventional methods for utilisation of water such as through inter-basin transfers, artificial recharge of ground water and desalination of brackish or sea water as well as traditional water conservation practices like rainwater harvesting, including roof-top rainwater harvesting, need to be practiced to further increase the utilisable water resources". This would suggest encouragement for states to develop policies based on the more local traditional approaches that had served communities well in the past.

Centralisation *versus* decentralisation was the subject of a workshop at World Water Week in Stockholm in August 2006 [11]. Most of the presentations were examples of successful decentralisation, with limited coverage of centralised examples. This may be a true reflection of a world trend but with a small sample of cases this cannot be verified. The workshop concluded that there is a trend towards decentralisation with examples of good progress worldwide. One major obstacle to success in decentralisation occurs when the financial aspects have not been included. In other words, decentralisation of responsibility with centralised constraints on finance and cost recovery will not work. Another significant obstacle is a lack of suitably trained staff to assume management and technical responsibilities. This relates partly to the issue of scale discussed above and partly due to the need for sufficient time for the necessary skills to be acquired locally. Successful decentralisation requires transparency and accountability, providing benefits of ownership, empowerment, integration and affordability. This conclusion raised the very important point that a move to local government control can result in less transparency. Transparency is one of the key aspects for success raised time and time again in this book. The debate at the Workshop also covered the conflict between decentralization, necessarily resulting in smaller policy and operational units, and the need for integrated water resource management. This suggests that there is a need for centralised broad policy making, with local policy interpretation and delivery. A recent study in India [12] concluded that "decentralisation has a negative effect on production efficiency and that 'closer to the people' may not necessarily be better suited to provide water supply services". The main limitation here was seen to be poorer human resources development with the decentralised operating units neglecting operator training. My experience in India relates to the importance of planning and technical support to decentralised operational units. This is discussed in Chapter 3.

The benefits and limitations of both centralisation and decentralisation of water matters will continue to be debated. There is not a perfect model because the decisions depend not just on organisational theory, but also upon local history and whether or not a current situation is effective. This issue leads naturally to two other subjects, namely water resource planning and vertical integration.

WATER RESOURCE PLANNING

There is a tendency for water resource planning to be seen as Integrated Water Resource Management (IWRM), with emphasis on river basin management and the question of how a river's water is shared in a sustainable way between people living along its length and the allocation between cities and agriculture. At the World Summit on Sustainable Development held in Johannesburg in 2002, the Summit called on all countries to develop integrated water resources management and water efficiency plans by 2005. This was a laudable aim directed at the Millennium Development Goal of "Ensuring environmental sustainability". It was also very challenging, and the 2nd World Water Development Report 2006 [13] gave progress towards that 2005 target. The figures are reproduced below in Table 1.1.1. Clearly, the level of progress at that time was disappointing, although given the short time frame, not surprising. The report points to most progress being made where there had been a specific need to be addressed. I think this is important because people come together most effectively when working on a project with a defined practical output. It is for this reason that I like "drinking water safety plans" which bring together all the organisations involved from catchment to tap; I cover this subject in Chapter 4.

Table 1.1.1 Progress towards meeting the Johannesburg Target of IWRM Plans.

Region	Countries surveyed	Good progress	Some steps	Initial stage
Africa	45	6	17	22
Asia and the Pacific	41	5	20	16
Latin America	22	3	14	5
Total	108	14	51	43

Source: 2nd World Water Development Report (2006).

A later report, by UN Water in 2008 [14] suggested that progress had accelerated over an 18 month period, see Table 1.1.2. There is some doubt over the accuracy of these figures as they refer to data from two different monitoring organisations, namely The Global Water Partnership (GWP) and UN Water. My concern about the IWRM plan target is that because it was set for 2005, it was likely to have been seen as of higher priority by governments than the 2015 water supply and sanitation targets. I recognise that focussing on the water and sanitation targets alone would have been sub-optimal, but given the need for a strong bottom-up approach (discussed above) to meet those demanding targets, to paraphrase Voltaire "the best must not be the enemy of the good".

Table 1.1.2 Accelerated progress on IWRM Goals.

		Plans completed or being implemented (%)	
	Number of countries	**End of 2005 (GWP)**	**2007 (UN Water)**
Africa	24	25	38
Americas	14	7	43
Asia	15	27	33
Total	53	21	38

Source: UN Water (2008).

There are many references on the subject (see, e.g. [15], [16] and [17]). The book by The Global Water Partnership *The Boldness of Small Steps* [17] was published in August 2006 to denote the important 10-year landmark of the work of the organisation. Clearly, the concept and principles of IWRM are highly important, and the challenge is to find an effective approach to implementation, in which some necessary emphasis is given to IWRM in the context of delivering water and sanitation services. Below, I give a brief description of two examples, one urban and one rural, of successful integrated water resource planning, which have delivered sustainable water supplies. They are both in Asia, but wholly different in approach.

The first example is Singapore [18, 19] which has taken an integrated approach to water resource management utilising all sources of available water. In order to address major problems of flooding and pollution, the Singapore Government re-organised water and wastewater management by bringing the two parts together under the Public Utilities Board (PUB). PUB's task was to deal with the problems and provide Singapore with a sustainable water supply. On water source availability there is a concern over whether imported water from Malaysia will continue to be available after the end dates of the two agreements. The first agreement ended in 2011. The approach taken by the Government and PUB to meet their requirements has involved a 20-year clean up of the river, the development of multiple water resources including reuse, effective water conservation and a high level of leakage control. The so-called "4 Taps" approach to water sources is a combination of imported water from Malaysia, maximising the catchment collection within Singapore's own territory (including the construction of a new reservoir within the city by creating a freshwater marina), reuse of wastewater and desalination. The city-based sources include the collection of stormwater. Effective water conservation has been achieved through an innovative programme of consumer participation with the latest campaign Water Efficient Homes, to help residents save water at home and cut down on their water bills. It is called a PPP (Public, Private, People) approach, but it is a community driven one, with local organisations encouraging residents to have water efficient homes through the use of water saving devices and through water saving practice. One part of the campaign is making available "Do-it-yourself" water saving kits, which consist of spray thimbles for taps and water saving bags for toilet cisterns. PUB has set up mobile exhibitions to give advice to the consumers on good installation practice, and to show what can be achieved with water saving devices. Conservation measures include a water conservation tax being added to already full-cost recovery tariffs. By 1990, non-revenue water (or unaccounted for water, UFW) had been reduced to 10%, and through continuing effort, by 2006 it was down to 5%, largely due to leakage control, as the other aspects of the water balance had been dealt with earlier. Singapore is an excellent example of what can be achieved with clear and focused Government policy and separation of delivery under high quality management. Importantly, Singapore has adopted a policy of continuous improvement with development continuing into the next phase, which Singapore is calling "Our blue waterways". This ABC water programme, Active, Beautiful and Clean, involves integrating the water aspects with the wider environmental programme in Singapore. The ABC labelling again demonstrates the recognition of effective communication with the public and the need for active public buy-in and participation. A comprehensive account of the Singapore Water Story was published in March 2013 [20].

The second example [21, 22] is from rural China in the dry mountainous region of the Loess Plateau in Gansu Province. The area is at 800–1400 metres altitude in an area where there has been land erosion and consequent landslides. It is one of the poorest parts of China. The aim of the plan, set up in 1983, was to improve water and food supplies and to develop better farming practice. A rainwater harvesting approach was taken to provide water for domestic consumption and to provide water for irrigation. It involved simple innovative approaches in both water management and in farming practice. In 1995 and 1996 the local government provided financial support for materials to allow families to build their own collection and storage systems. The scheme provided 10 litres per day per person for 1.3 million people and 30 litres a day per animal for 1.2 million livestock. This is low by most standards but much better than before the schemes were introduced. It should be sustainable as families have the incentive to maintain their own systems. In 1997 rainwater harvesting on a catchment basis was added to provide supplemental irrigation.

VERTICAL INTEGRATION

I believe that the most desirable form of vertical integration is to have all aspects of water supply (see Figure 1.1.1), plus the equivalent parts of the wastewater system, managed by the same organisation. The Singapore example of integrated water resources given in the last section is a perfect illustration of the value of having such vertical integration. Others who believe that there are benefits from direct competition will disagree. It is perhaps a question as to whether increased interface costs between players exceed the trading benefits. A section on competition is given in Chapter 9.

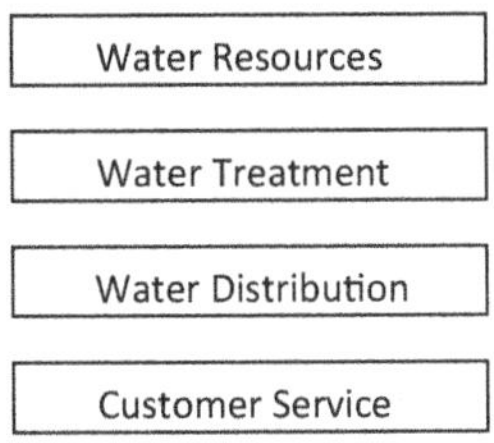

Figure 1.1.1 Vertical integration of water supply.

Integration of all four parts of water supply is important, as it provides not only for the important hydraulic and water quality linkages, but also for the vital relationship between costs, revenue collection and customer service. It is particularly important for the integrated management of water treatment, water distribution and customer service. The effect of the lack of such integration is evident in BOT (Build, Operate and Transfer) projects which separate treatment from distribution and customer service and, in some cases, limit the ability of management to optimise the use of water from other treatment works which is supplied through the same distribution system. A BOT contractor needs to be confident that there will be an income stream for the duration of the contract, before being willing to commit money to the construction of the plant. This has resulted in contracts in which the utility has been committed to paying for a volume of water, quite independently of the demand or of the supply circumstances at its own treatment plants. I give examples of such cases in Chapter 8.

There is also the question of the cost of drinking water quality monitoring, with the need for the quality of the water from all sources entering the distribution system to be monitored to reduce the risk of contaminated water reaching consumers. If treatment and distribution are managed separately, monitoring becomes not only a technical requirement, but also a contractual issue. This becomes critical in countries like the UK where the supply of drinking water "unfit for human consumption" is a criminal offence. I will be interested to observe what monitoring requirements become necessary with common carriage. The emerging scene in England is discussed in Chapter 9.

THE IMPORTANCE OF CONTINUOUS SUPPLIES (OFTEN REFERRED TO AS 24/7)

Those of us living in developed countries have become accustomed to having water at our taps at adequate pressure 24 hours a day, every day. We complain if a supply is interrupted due to a system failure or for planned maintenance, even though it does act as a reminder of how lucky we are to have such a good service. Others in the world are less fortunate, and either don't have a convenient tap supply at all, or have a supply for a few hours a day, and not every day. To my dismay, in my discussions with some government officials and some water managers in those parts of the world, they see intermittent supplies as necessary and inevitable. Firstly let's consider the conditions that result in intermittent supplies. Either there is insufficient water to meet demand, or too much of it is leaking out of the pipes, or the distribution system does not have the capacity to deliver the water. Taking each of these in turn, demand may have outstripped supply because of a growing population or because of increased per capita needs for water, or because water is being wasted. Wastage can result from consumers feeling the need to take more water than required during the periods when there is a water supply, not knowing when it will next be available. A policy of intermittent supplies can be argued as necessary to reduce leakage, but leakage levels as high as 60% have been estimated in some such areas [23]. The effect of intermittent supplies, with fluctuating pressure in distribution systems could accelerate the deterioration of pipes thus making leakage worse. Measurements in Cyprus [24] during a short period of non-continuous supplies showed increased leakage levels and an increased number of burst pipes. Also, high leakage limits the effective capacity of distribution systems and requires higher pressures at source to deliver the same amount of water to consumers. It may be necessary to increase water resource capacity, but it is not possible to deliver increased treated water to consumers through a leaky distribution system. It is a salutary thought to recognise that, with 60% leakage, the amount of water required to be put into supply to provide the consumers with the 40% they are receiving is 2.5 times as much than if there were no leakage, and twice as much than if leakage were 20%. Even if leakage is at an "acceptable" level, a distribution system may not have the capacity to deliver water to all the consumers connected to the system. This can be because a system has been extended beyond its original design capacity, or because in the case of cast iron systems, corrosion products have reduced the diameter of the pipes, or both.

I can quite understand the attitude of those having to manage intermittent supplies, as the challenge of recovering from the set of circumstances outlined above must be daunting, but so is the prospect of continuing to live with the current low level of supply, with the associated risk of ingress of contaminated water from the ground when there is no pressure in the systems. This is not a technical problem because the necessary elements of network analysis, system pressure control, leakage control and distribution system rehabilitation can easily be put in place. It is generally regarded as a financial problem with money not being available for this purpose. On the face of it, it is a finance question but the real problems are far more fundamental. They relate to a lack of critical elements of policy, particularly those of sustainable cost recovery, management performance measurement, and a lack of public participation and transparency. Without adequate income it is not possible to operate and maintain a system effectively, nor is it possible to attract competent management. Without effective public participation it is not possible for consumers to understand that low tariffs are not in their best interests. Of course, the first steps are difficult because of a backlog of remedial work and insufficient income to pay for it. Consumers are naturally wary about paying more unless they are confident that the service will improve; see Figure 3.1 on Page 40. It requires Government to recognise the need, establish the necessary direction and policies, and then seek the means of financing the gap in income until tariffs are at a sustainable level. Phnom Penh is a good model of how sustainable and inclusive 24/7 can be achieved. The successful development is described in Chapter 7.

In India the transition has started. Three 24/7 pilot projects which were completed in 2010 in the State of Karnataka, India, demonstrated the importance of 24/7 in managing water resources and in providing continuous sustainable water services for all [25]. Jalakam [26] in discussing the results of the continuous supply demonstration projects in Karnataka concluded: "the results have clearly demonstrated that providing continuous water supply does not require more water resources but necessitates an efficient distribution network and a prudent management practice of water loss control." One hundred per cent coverage was achieved in the 24/7 project areas and it was found that the poor, many of whom were previously not served by the network, were willing and able to pay for a continuous service. As a result of the success of the 24/7 projects in Karnataka, 24/7 is now the Indian national service level benchmark and the basis for estimating investment needs. After Delhi's earlier and bitter opposition to 24/7 in the belief that it was part of a move to privatisation, its civil society groups are now welcoming the proposed implementation [27].

The experience in India is showing that it should be possible to make rapid progress. In others, with less money for investment in the infrastructure, it will require external financial assistance during the transitional period. To my mind donor investment in the development and rehabilitation of distribution networks would yield far greater benefit in the shorter term, than investment in additional water resources, although in some areas both may be required. In this section the need for sustainable cost recovery has been discussed in the context of intermittent supplies. Intermittent supplies are one symptom of the failure of governments to introduce appropriate water charging policies. This fundamental requirement of sustainable cost recovery appears throughout the book.

SANITATION POLICY

If water supply policy has been neglected in some countries, the situation with regard to sanitation is far worse. When introducing piped water supplies, it is essential that consideration is given to how the used water will be handled. This means having policy and plans on sanitation that are appropriate to an area. In the developed world we have become accustomed to having a waterborne system in which we flush and forget. Such systems are attractive to the public, but they are expensive, they require a lot of water, and they often result in high volumes of treated, or sometimes untreated, water being discharged at one point into water courses. Generally that does not matter because great attention has been given to the conditions for those discharges. By and large, the developed world is no longer worrying about point sources on environmental grounds, but has moved on to having concerns about diffuse pollution.

Different approaches have been developed for sanitation, one being the design of dry or semi-dry systems requiring little or no water. These so-called ecosan systems have been very effective in areas in which previously there was no dignified sanitation, and where there was sufficient land for local handling of both the liquid and solid content from the toilets. Between the fully waterborne systems and the fully dry approaches, there is a range of systems which take some elements from each end of the spectrum, and each one of them has a part to play in providing dignified sanitation for all. The important policy question is how to decide which system is the most appropriate for a particular area. The answer is not to listen to an advocate of one particular approach or another, as that is likely to lead to the wrong answer. What is needed is an objective way of evaluating needs and means, area by area.

Following concern at the lack of good information to allow objective debates on sanitation at the UN meetings on Sustainable Development, at the World Water Forum and elsewhere, the International Water Association (IWA) formed a Task Force of experts to develop a coherent structure for the classification of systems, and to provide an approach to assist policy makers and planners in their decisions on sanitation. The project was called Sanitation 21 and the Task Force launched its first output, called *A Draft Framework for Analysis* [28] at a Workshop held at the IWA Congress in Beijing in September 2006. The Task Force argued, and I quote, "improving the quality and effectiveness of sanitation investments is not particularly about technologies, ... rather it is about developing an explicit understanding of what the objectives of a system are and then designing a system that meets those objectives", and goes on to say, "all we are really saying is let's do planning and design better". This sounds simple but achieving it is difficult because each part and level of a community has a different perspective on objectives. Recognising those differences is an important starting point. Figure 1.1.2, which is taken from the DFID Guidance Manual on Water Supply and Sanitation [29] provides a good illustration of the community levels or domains, each having its own objectives, and each level needing to be involved in the planning decision process. The completed work resulted in planning guidance which is freely accessible on the IWA website [30] together with a number of associated references.

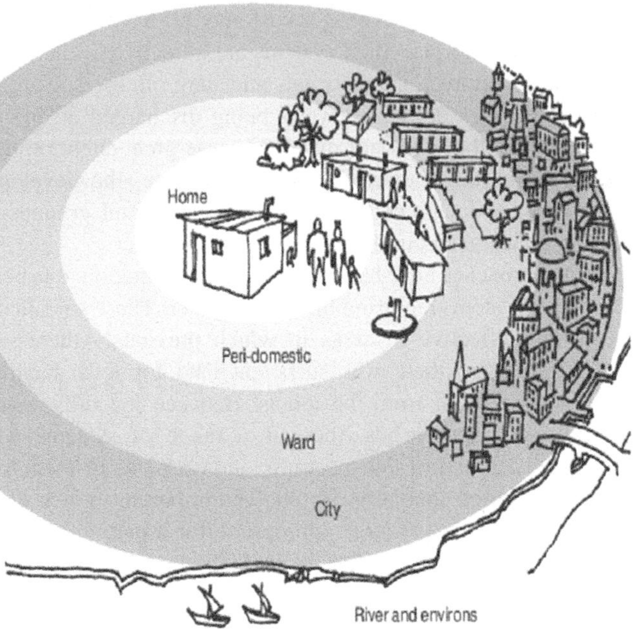

Figure 1.1.2 Levels of perspective on sanitation objectives (*Source*: DFID guidance manual on water supply and sanitation programmes [29]).

Sanitation 21 has brought some structure into the decision making process by providing a useful classification of the technical options. The sanitation options are listed as three main systems with a number of sub-options:

(1) Dry sanitation
  A: On site dry
  B: On site dry with semi-centralised treatment
  C: On site dry with urine diversion.
(2) On site semi-wet sanitation
  A: On site semi-wet (pour-flush)
  B: On site wet with (semi) centralised treatment.
(3) Waterborne sanitation
  A: Waterborne with (pre) treatment and (semi) centralised treatment
  B: Waterborne with (semi) centralised treatment
  C: Waterborne with centralised treatment.

The challenge for governments is to establish good planning systems which allow for understanding of the interests and incentives across all domains of the city, to understand the constraints due to water and land availability, costs and environmental considerations, and to maintain objectivity in making decisions. In particular, it is necessary to avoid following any particular sanitation fad or fashion. Sanitation 21 gives a number of case studies, which are valuable in illustrating the strength of the approach and the need for effective governance and regulation.

CORRUPTION

Transparency International (TI) [31, 32] is in the forefront of efforts to tackle the problem of corruption which it defines as "the abuse of entrusted power for private gain". TI estimates that up to 40% of international funding is "lost" within governments even before the procurement stage, and that corruption can add up to 25% to the cost of public contracting. Jamal Saghir, in his keynote address [33] at the International Water Association Congress in Beijing 2006, said that in public contracts there can be 15% excessive cost due to collusion between contractors and 6–11% added to the contract value due to kick-backs. There are cases where 41% of consumers have paid bribes to meter readers to falsify meter readings and 20% of consumers have paid bribes to obtain illegal connections. A survey [13] carried out in India in 2004, gave a good indication of the frequency of kick-backs from contractors to public officials. The information on kick-backs gleaned from public officials who took part in the survey was as follows

- 17% said that it takes place every time;
- 33% said that it was quite common;
- 8% said that it takes place about half of the time;
- 17% said that it occurs occasionally;
- 25% said that it occurs infrequently/never.

This is a gloomy picture, and it leaves one in no doubt that corruption needs to be taken seriously as it undermines public services and public confidence. Transparency International has produced guidance, *A Toolkit for Transparency in Local Governance* [34], with the aim of assisting the process to "empower citizens to work for a world free of corruption". It gives an approach to "municipal corruption vulnerability assessment", provides advice on public participation and on access to information laws and media training. To my mind it should be required reading for all those involved in policy and for NGOs seeking to achieve greater transparency.

There is also a fundamental problem to be addressed: that of the remuneration of public service employees. It is unrealistic to expect those working in public service to resist the temptation of bribes when their salaries alone are insufficient to provide them with an acceptable standard of living. This question of public service pay is associated with other aspects of water service management, including that of having sustainable cost recovery as part of a stable financial structure. These financial issues need to be addressed in the round to establish an environment which will attract good management and staff, which in turn will lead to more-efficient operations. Public sector reform is discussed in Chapter 6.

At a meeting held in Delft, The Netherlands in November 2005, a number of organisations involved in the water sector met with Transparency International, and decided to set up a Water Integrity Network (WIN) aimed at promoting transparency and integrity as well as combating corruption in the water sector. WIN was established in 2006 and in its literature it says that "it is open, inclusive and promotes solutions-oriented action and coalition-building between civil society, the private and public sectors, media and governments. WIN advocates a wide range of anti-corruption activities locally, nationally and globally." WIN, with active participation from the Stockholm International Water Institute (SIWI), has produced a document [35] entitled *Corruption in the Water Sector, Causes, Consequences and Potential Reform.* This document gives some key recommendations on addressing corruption, which include recognizing that corruption is a symptom of established social systems needing reform, and that anti-corruption measures should be aligned with national governance reform. It talks about anti-corruption measures being in uncharted territory and that there is a need for patience, resources and political and institutional reserves. WIN and Transparency International have produced a manual [36] to assist in overcoming corruption in public contracts. It advocates an "integrity pact" approach. One aspect is the use of monitors who at each stage, from tendering and award of contract to contract completion, check for indications of corruption. I quote from the manual – "During the contract execution stage, most corruption risks are associated with bribery and kickbacks to secure positive audit and oversight reports, so it is good to have a third party watching". It is stressed that monitors should be independent, knowledgeable, have an ethical reputation, be accountable, and have capacity and commitment. NGOs are included in those considered suitable for the monitoring role. Chapter 6 is devoted to the effective use of the private sector and I include a discussion on the benefit of having an independent regulator who can oversee the contracting process. A regulator should not interfere in the contract bids but ensure that the tender, and any later reviews, are based on realistic expectations and sound information.

CONSUMER PARTICIPATION

Consumer participation is much more than a means of combating corruption. It has become a key aspect in the development of sustainable water services. At one time, consumers trusted their governments and water professionals to take decisions in their best interests. With better education, and with the development of "consumer power", this is no longer the case. Consumers want to be involved, or at least to know that there is transparency and that they can be involved if they have concerns. The level of involvement is an important policy matter and can range from consumers simply receiving information to the establishment of citizen control. The policy question for national governments is how to achieve effective public involvement. This question is covered in Chapter 5, Transparency and Public Participation, which gives examples of successful approaches.

COST RECOVERY, TARIFFS AND AFFORDABILITY

This subject is covered in detail in Chapter 3, but it is important to cover briefly some fundamentals in this chapter. I subtitled this chapter "Some bricks in the wall". Without this brick, the wall will collapse. I include this here because there is more nonsense talked about it than almost anything else. In the developing world the poor are paying water vendors anything up to 25 times more for a litre of water than those who have a subsidised tap water supply. [37] What the poor cannot do is save money to pay for an access charge or to pay monthly bills. They need a means of paying an access charge in easy instalments and of paying for water "as they go". The lack of tap supplies for the poor is associated with low (i.e. below cost recovery) water tariffs, which result in the run-down of current systems and the inability to finance extensions and improvements. Full cost recovery is essential for sustainability and should be part of policy for making effective provisions for the poor. Increasing charges is unpopular, and moving from a current low tariff situation requires political courage. Involving citizens so that they are able to understand the need for increased charges, in the provision of a better service, is a good starting point.

WATER FOR HEALTH

Generally, discussions about drinking water relate to water for survival or concerns over water quality, but there is another important aspect, that of hydration. With the availability of coffee, tea and commercial soft drinks, there is a tendency for individuals to think that they do not need to drink water. The result is a poor level of hydration with adverse implications for health. Water UK, the Association of UK water and wastewater service suppliers, set up the Water for Health Alliance in 2002. The starting point was to commission a review [35] of scientific literature on the impacts of water (or lack of it) on health. The review confirmed the importance of drinking a sufficient quantity, around 2 litres per day for an average adult, to reduce the likelihood of developing certain cancers, in controlling hypertension, preventing or reducing other medical conditions, and in improving general wellbeing. As is well known, full hydration is essential for athletes. It is also important in the effectiveness of medication, especially in the elderly. Some commercial soft drinks have been shown to make children more disruptive, whereas water improves concentration at school. As a result of the

Alliance's work, the provision of water is now becoming an integral part of government's and others' thinking on policy in institutions throughout the UK. On the Water UK website [39] there are the following documents giving advice on hydration: Hydration Best Practice Toolkit for Hospitals and Healthcare; Water for healthy ageing (toolkit); Wise up on water! (children, older people, cancer prevention – 3 factsheets) and Water in the workplace (factsheet). I see this aspect as important for public health, and also in helping to give recognition to the valuable product that is delivered to consumers' taps at a cost of around 0.1 UK pence (0.15 US cents) per litre. Whereas the focus has to be on the many problems with water safety in many parts of the world it is good to reflect on the importance of hydration to health.

NATIONAL WATER AND SANITATION PLANS

The extremely important *Human Development Report 2006* [1], produced by UNDP, with its subtitle *Beyond Scarcity – Power, Poverty and the Global Water Crisis*, stresses the need for political leadership on water and sanitation, and the need for the establishment of "dedicated water and sanitation ministries led by senior cabinet ministers". It says that "each country should have a national water and sanitation plan, integrated in national poverty reduction strategies, and reflected in medium-term financing frameworks and budgeting priorities". This chapter has not been written to give the structure for such plans, but each of the subjects covered needs to be considered in the development of plans. I hope that the book will be of assistance to those who have the task of preparing water and sanitation strategies plans.

REFERENCES

1. Governance for Sustainable Human Development (1997). A UNDP Policy Document. United Nations Development Programme January 1997, New York.
2. UN (United Nations) (2011). World Population Prospects: The 2010 Revision. United Nations, New York.
3. Bartram J., Elliott M. and Chuang P. (2012). Getting wet, healthy and clean: why households matter. *Lancet*, **380**, 85–86.
4. UNICEF and World Health Organisation (2012). Progress on Drinking Water and Sanitation. 2012 Update, http://www.unicef.org/media/files/JMPreport2012.pdf (accessed May 2013)
5. Resolution adopted by the UN General Assembly 64/292 (2010). The Human Right to Water and Sanitation. 3 August 2010.
6. Bridging the Gap (2006). Citizens Action for Accountability in Water and Sanitation. March 2006.
7. Schienle S. (2004). Self-administration by users in the delivery of water and wastewater services. The Systema Integrado de Saneamento Rural, SISAR, Stockholm Water Symposium, August 2004, Stockholm.

8. Okun D. (1977). Regionalisation of Water Management, A Revolution in England and Wales. Applied Science Publishers, UK.
9. The Water Framework Directive (2000/60/EC) December 2000.
10. National Water Policy (2002). Ministry of Water Resources. Government of India, April 2002, New Delhi.
11. World Water Week in Stockholm (August 2006). Proceedings Stockholm International Water Institute.
12. Asthana Anand N. (2012). Decentralisation, HRD and production efficiency of water utilities; evidence from India. *Water Policy*, **14**, 112–126.
13. Water – A Shared Responsibility: Second World Water Development Report 2006. http://www.unesco.org/new/en/naturalsciences/environment/water/wwap/wwdr/wwdr2-2006/ (accessed June 2013)
14. Status Report on Integrated Water Resource Management and Water Efficiency Plans. Prepared for the 16th Session of the Commission on Sustainable Development – May 2008. UN Water.
15. van Hofwegen P. J. M. and Jaspers F. G. W. (1999). Analytical Framework for Integrated Water Resources Management. Published by Taylor & Francis, UK, Jan 1999. ISBN: 9789054104735.
16. Pangare V., Pangare G., Shah V., Neupane B. R. and Somasekhar P. (2006). Global Perspectives on Integrated Water Resources Management, 2006 edition. Published by the Academic Foundation. ISBN: 9788171885619.
17. The Boldness of Small Steps, The Global Water Partnership, August 2006. http://www.gwp.org/Global/The%20Challenge/Resource%20material/10th_anniv-en.pdf (accessed June 2013)
18. Khoo Teng Chye (2006). Keynote Presentation Stockholm Water Week August 2006.
19. Tortajada Cecilia (2006). Water Management in Singapore. *International Journal of Water Resources Development*, **22**(2), 227–240, June 2006.
20. Tortajada C., Joshi Y. and Biswas A. (2013). The Singapore Water Story: Sustainable Development in an Urban City State. Routledge, London.
21. Zhu Qiang and Li Yuanhong (2006). Integrated Rural Development with the Rainwater Harvesting Approach. Rainwater Harvesting and Management Workshop. IWA Water Water Congress. Beijing, September 2006.
22. Zhu Qiang and Liu Changming B. (1998). Rainwater utilization as sustainable development of water resources in China. Presentation at the World Water Week, Stockholm, 1998.
23. 'Billions of Dollars Down the Drain in Thirsty Cities'. Interview with Kaj Bärlund, the Director of the Environment and Human Settlements. UN-ECE, May 2006.
24. Water Loss Newsletter No 2. Management of Water Losses in a Drinking Water Supply System. Water Board of Nicosia, September 2011.
25. WSP (2010). The Karnataka Urban Water Sector Improvement Project. 24/7 Water Supply is Achievable. Water and Sanitation Program, Field Note, September 2010. Water and Sanitation Program, New Delhi.
26. Jalakam A. (2012). Continuous Water Supply – An Essential Component for Achieving Water Security. Available at: http:www.water.ox.ac.uk120416conference10a-jalakam.pdf (accessed May 2013)

27. Asthana V. (2008). Discourses of Power and Resistance in the Water Policy Process of Delhi, India. Unpublished PhD thesis, University of Illinois, Urbana, Illinois.
28. Sanitation 21 (2006). Simple Approaches to Complex Sanitation. A Draft Framework for Analysis. IWA, September 2006, The Hague.
29. DFID Guidance Manual on Water Supply and Sanitation Programmes (1999). Department for International Development, London.
30. Sanitation 21 Guidance. http://www.iwahq.org/Home/Development/Technical_Expertise/Sanitation_21/ (accessed March 2010)
31. Transparency International Annual Report 2004. http://www.transparency.org
32. Global Corruption Report: Corruption and Health (2006). Transparency International 2006. Pluto Press, London, Ann Arbor.
33. Jamal Saghir (2006). World Bank. Keynote Address, IWA Congress Beijing 2006.
34. Tools to Support Transparency in Local Government. Transparency International and UN Habitat, March 2004.
35. Stålgren P. (2006). Corruption in the Water Sector: Causes, Consequences and Potential Reform. Swedish Water House Policy Brief Number 4 SIWI 2006.
36. Integrity pacts in the Water Sector. Water Integrity Network & Transparency International 2010. http://www.transparency.org
37. Sharma S. (2002). Water Markets Exclude the Poor. World Summit on Sustainable Development, Johannesburg.
38. Forrester H. (2002). Water for Health. A Report of the Published Literature on the Impacts of Water on Health. Water UK, 2002.
39. Water UK publications on Water for Health. http://www.water.org.uk/home/saearch (accessed June 2013)

Chapter 1

Part 2 – Water and sanitation as a human right

Universal Coverage or a Charter for Lawyers!

Kofi Annan, former UN Secretary General, in his message for World Water Day, 22 March 2006, said "Access to safe water is a fundamental human need and, therefore, a basic human right. Contaminated water jeopardizes both the physical and social health of all people. It is an affront to human dignity." No-one can have any difficulty with that statement. It is absolutely fundamental and should be a key statement in every country's water policy. Four years later in 2010 there was the UN resolution [1] which recognised both water and sanitation as a human right.

It would be a mistake to regard the attainment of the UN resolution as an end in itself. It can be seen as a big step forward in that it gives water and sanitation political status, but the realisation of human rights is a highly demanding task [2]. Success in implementation will depend not just on States and political will, but on ordinary citizens accepting responsibilities to go with the rights, and their willingness to help protect the weak and vulnerable. If a wholly legally-driven approach to implementation were to be allowed it would result in chaos, but I believe it is an opportunity to accelerate progress towards universal coverage, providing it is implemented in a practical way. My reading of the text of the resolution is that it was written with recognition of the challenges and potential pitfalls.

There are concerns that the environment will lose out if priority is given to human beings. For example Bakker [3] writes "the human right to water (and various forms of community water-supply management) address redistributive concerns in economic and social terms, but overlook important environmental aspects of the urban water crisis: water scarcity, threats to water quality and ecological sustainability". I don't believe that it will mean that the environment will be ignored because concern for ecology should be part of addressing the declaration element of sustainability. Experience [4] has shown that long-term economic development, with water services for all, requires a sustainable development approach. A joint report [5] from OHCHR and UNEP, Human Rights and the Environment Rio+20, stresses the importance of an economic system "that recognizes the properties of healthy ecosystems as the backbone of economic and social well-being and as a precondition for poverty reduction." Water and sanitation are key components of poverty reduction.

To a large extent the elements for sustainable water services covered in this book are those which have to be addressed in meeting the so-called normative and cross-cutting elements in the resolution.

HISTORY

Although concepts of human rights had been debated as far back as the Middle Ages, there wasn't an international vehicle for promoting universal human rights until the formation of the United Nations in 1948. On December 10, 1948, the Universal Declaration of Human Rights (UDHR) [6] was adopted and proclaimed by the United Nations General Assembly. The UDHR covered two areas of human rights, civil and political rights, and economic, social and cultural rights which included health. It took until 1976 before most countries were willing to adopt treaties to give legal force to both areas in the form of covenants. In 2002, the UN defined the extent and scope of the human right to water in General Comment 15 [7]. It read "The human right to water entitles everyone, without discrimination, to sufficient, safe, acceptable, physically accessible and affordable water for personal and domestic use." This was opposed by a number of countries, including the United States, Canada and Mexico. In my view, these countries agreed with the text as a principle but were fearful that it would lead to legal demands for water services to be free. Although there were active campaigns it wasn't until after 2010 that the reluctant countries became willing to adopt water and sanitation as human rights. In my view again, the political change occurred because countries recognised that the approach taken by the UN, as a result of well-developed proposals by the UN's Independent Expert, could be implemented in a practical way. A timeline of stages towards the UN resolution is shown in Figure 1.2.1. This includes various attempts to improve access. There was the International Decade for Water and Sanitation, 1980–1990, and probably the most important initiative before the UN resolution that of the Millenium Development Goals (MDGs). As I write decisions are being taken on what approach will be taken post 2015. It can be expected that the next stage will be influenced strongly by the elements in the UN resolution which are discussed later.

The key appointment of the Independent Expert is shown on the timeline but one important date not included is the step taken by South Africa in 1996 to include a right to water in its Constitution [9]. This was a significant step for South Africa but also for the world as it gave practical focus to water as a human right. Clause 27.1(b) of the Constitution states that "everyone has a right to sufficient food and water". This was followed in 1997 by the Water Services Act [10], Section 3 of which states that:

(1) Everyone has a right of access to basic water supply and basic sanitation.
(2) Every water services institution must take reasonable measures to realise these rights.
(3) Every water services authority must, in its water services development plan, provide for measures to realise these rights.

Increasingly other countries are introducing legislation. The list of countries is kept up-to-date by the Information Portal on the Human Rights to Water and Sanitation [11].

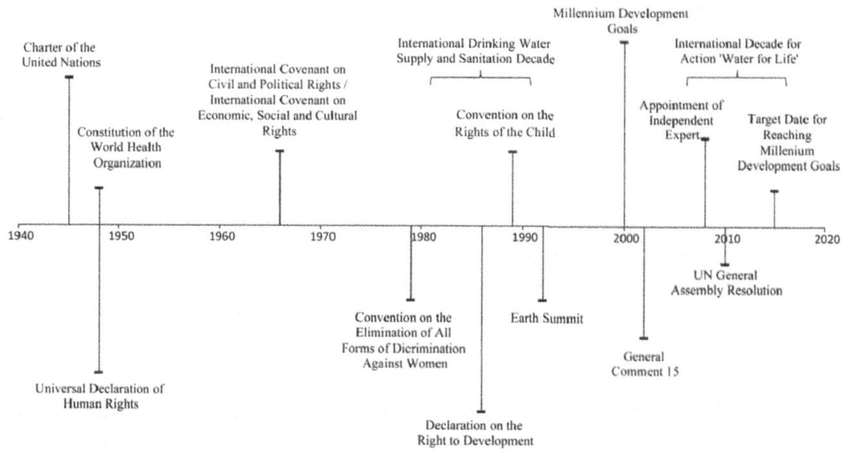

Figure 1.2.1 Key steps in pursuit of universal access to water and sanitation (*Source*: Onda, K., LoBuglio, N. and Bartram, J. (2012) [8]).

THE UN RESOLUTION

The resolution [1] reads as follows:

(1) *Recognizes* the right to safe and clean drinking water and sanitation as a human right that is essential for the full enjoyment of life and all human rights;
(2) *Calls upon* States and international organizations to provide financial resources, capacity-building and technology transfer, through international assistance and cooperation, in particular to developing countries, in order to scale up efforts to provide safe, clean, accessible and affordable drinking water and sanitation for all;

(3) *Welcomes* the decision by the Human Rights Council to request that the independent expert on human rights obligations related to access to safe drinking water and sanitation submit an annual report to the General Assembly, 13 and encourages her to continue working on all aspects of her mandate and, in consultation with all relevant United Nations agencies, funds and programmes, to include in her report to the Assembly, at its sixty-sixth session, the principal challenges related to the realization of the human right to safe and clean drinking water and sanitation and their impact on the achievement of the Millennium Development Goals.

The substantive document on which the resolution was based was the Independent Expert's Report [12] to the UN General Assembly dated 3 August 2010. This includes the key normative and cross-cutting elements which "define" the human rights to water and sanitation, which are listed below. A book [13] on good practice followed in March 2012.

THE ELEMENTS

The elements in the Independent Expert's Report provide the framework for implementation. Extensive work is taking place within a UN framework with involvement by the International Water Association (a network of water professionals) to produce implementation handbooks, which should be available in draft at the end of 2013 for review worldwide and for general use by the end of 2014. The handbooks, which will be based on good experience in achieving universal access, should provide the best chance of successful implementation of the resolution. An important aspect will be on avoiding pitfalls. This chapter does not attempt to provide interpretation of the elements or guidance on implementation, but does present my views on some of the issues which have to be addressed. To set the scene I reproduce below key sentences from the Independent Expert's documents ([7], [13] and [14] without comment.

Normative elements – water

- *Availability*: Water must be available in sufficient quantity to meet requirements of drinking and personal hygiene, as well as further personal and domestic uses, such as cooking and food preparation, dish and laundry washing as well as cleaning.
- *Quality/Safety*: Water should be of such a quality that it does not pose a threat to human health.
- *Acceptability*: Water should be of an acceptable colour, odour and taste.
- *Accessible*: Water has to be accessible. At the minimum, the human right to water requires that the distance to the water source should be in reach of

every household. Supply needs to be accessible on a reliable and continuous basis. Water must be supplied at work, school and in public places.

- *Affordable*: Use of water facilities and services must be at a price that is affordable to all people.

Normative elements – sanitation

- *Availability*: The criterion is related to the availability of facilities and services as well as the necessary structures to ensure availability of services.
- *Quality/Safety*: Human rights require that sanitation facilities must be hygienically safe to use and easy to clean. Regular cleaning, emptying of pits or other places that collect excreta, and maintenance are essential for ensuring the sustainability of sanitation facilities and continued access.
- *Acceptability*: Sanitation facilities and services must be culturally acceptable. In most cultures, toilets must be constructed so as to ensure privacy.
- *Accessibility*: Sanitation facilities must be physically accessible for everyone within, or in the immediate vicinity of each household, health or educational institution, public institutions and places, and the workplace. Sanitation facilities must be accessible on a reliable and continuous basis to satisfy all needs throughout the day. The sanitation facility itself should be secure and convenient for all users.
- *Affordable*: Use of sanitation facilities and services must be at a price that is affordable to all people.

Cross-cutting elements

- *Non-discrimination*: Water and sanitation must be provided without discrimination of any form, and particular care must be taken to provide service to those who are not able to provide for themselves, as well as to excluded individuals and groups and those at risk.
- *Participation*: All actions that have an impact on people's access to water and sanitation services must provide meaningful opportunities for engagement. Users, particularly those who are generally under-represented, including women, ethnic and racial minorities, and marginalised groups, must have an opportunity to participate in decision-making.
- *Accountability*: States are obliged to respect, protect and fulfil the rights to water and sanitation, and should be held accountable for meeting these obligations. Accountability .... will include monitoring, complaints mechanisms, dispute resolution and transparent governance.
- *Sustainability*: Practices should be economically, environmentally and socially sustainable so that future generations can enjoy rights to water and sanitation, and must therefore look beyond the short-term goal of assisting

people to gain access to water and sanitation services. Service providers and policy makers are therefore responsible for ensuring that this is integral to planning, and their responsibility does not end with the mere provision of a facility.

The above guidance does not amount to definitions of the elements; they are qualitative not quantitative but they do provide a framework on which implementation policies can be established. There is recognition of the scale of the challenges in achieving universal access, through use of the term "progressive realisation". What this means in practice is discussed later under the heading "International Targets and Monitoring".

SOME KEY ISSUES

Water services cannot be free: The resolution recognises that water and sanitation services have to be paid for. However, there is a risk that there will be legal attempts to obtain free services. Free services (See Chapter 3) are not sustainable for two reasons. Firstly consumers waste water if they don't have to pay for it, with examples of consumption having trebled. Secondly, general subsidies (which benefit the rich more than the poor) are neither sufficient nor reliable resulting in inadequate maintenance of infrastructure and a downward spiral of service. Sustainable cost recovery is necessary [15] through water and sanitation service charges to maintain facilities not only for use by future generations, but also to prevent loss of access by those currently being served. Loss of service is a loss of a human right and questions on non-regression in the level of service were the focus of the Independent Expert's questionnaire to States in 2013 [16]. Some emphasis was placed on refurbishment of existing infrastructure.

Those consumers, who are able to do so, have a responsibility to pay their water bills. This raises the question of redress for non-payment and the controversial issue of service cut-off. It is illegal in the UK for a service provider to cut-off a water supply. As a result non-payment in England (See Chapter 9) has become a serious problem. Affordability is an issue for the poor, although it can be more a question of payment methods (pay as you go) rather than absolute amounts. There are good examples of targeted subsidies for the poor, for example in Chile (See Chapter 8).

Existing infrastructure maintenance and refurbishment cannot be ignored: Sustainability is part of the cost recovery issue above, but it is also a major policy and engineering issue. A lack of maintenance and refurbishment result in high leakage losses. This in turn limits the ability to provide continuous supplies to existing consumers, and the introduction of hours a day or week rationing. This approach involves rotation in pressurising parts of the distribution systems. Depressurisation and repressurisation result in further deterioration in the structure of the pipes, and a consequent downward spiral in the service provided.

This problem is discussed in some depth in a section in Chapter 1 Part 1. Moreover, it has now been demonstrated that such a method of rationing requires more, not less, water resources. In urban areas, with piped supply systems, continuous (so-called 24/7) supplies will be essential to meet the requirement of "reliable and continuous service". Without a robust infrastructure it will not be possible to extend the service to those currently unserved [15].

Safe and acceptable drinking water cannot mean pristine: Drinking water must be safe. The primary requirement for safety is that it should be free from pathogens and toxic levels of chemicals. Although WHO guidelines provide the ultimate goal it is neither necessary nor affordable for that to be achieved in the shorter term. It should be recognised that many of the guideline values for chemicals are based on no adverse effect on a person weighing 60 kilogrammes drinking 2 litres per day for 70 years, and with a factor of safety, sometimes as high as 1000. Given limited finance, much greater impact will be made by achieving interim standards for everyone rather than WHO guideline quality for a few. Later, as economies grow as a result of improved water and sanitation, and thus improved health, the bar can be raised with progressive realisation towards universal WHO guideline quality. A (drinking) water safety plan approach will reduce risks, and provide an affordable way forward. See Chapter 4.

Are market-driven approaches to service delivery compatible with human rights? My plea here is that the implementation of human rights should not become a battle ground between those promoting the use of the private sector and those who are opposed [17] to private sector involvement. The requirement is that everyone who can further effective implementation is encouraged to be involved. The public/private argument has blighted progress with critical human endeavour resources having been expended on unproductive arguments. In the sub-heading I refer to market-driven rather than privatisation approaches because to my mind the main question is not one of ownership but of how to achieve efficient affordable services (See Chapter 6). The trend towards having publicly-owned "private" companies provides one effective model for progress. There have been failures of both public and private operations. Chapters 7 and 8 discuss reform of the public sector and effective use of the private sector respectively. Successful implementation of the UN resolution will require putting in place the elements of governance covered by the chapters in this book.

Minority groups cannot be given special priority, nor should they be at the end of the queue: They should be fitted into integrated planned phases of progressive realisation which make best use of available finance and water and sanitation infrastructure. Legal intervention to push particular cases could slow down overall progress.

Public Participation: Participation is given as an element in which all parts of the communities must be consulted and involved in policy. I go further than that and believe that without such participation it will not be possible to implement water and sanitation as a human right. There has to be public understanding of the timescales involved, of the need to use water wisely and of the need to pay for services. Public involvement in discussing and agreeing service development programmes will be paramount to understanding why everyone can't be at the front of the "queue". Approaches to public participation and successful examples are given in Chapter 5.

INTERNATIONAL TARGETS AND MONITORING

The scale of the task to achieve universal access to water and sanitation is immense. It has been estimated [18] that the worldwide infrastructure investment requirement for water and sanitation over the period 2005–2030 alone is US$23 trillion. This is greater than that required for energy and transport combined. The Joint Monitoring Programme Report (JMP) [19] states that, although there has been significant progress in extending access, in 2010 over 780 million people were still without access to improved sources of drinking water and 2.5 billion people still lacked improved sanitation. In addition the world's population is expanding rapidly with estimates [20] that the world's population will increase from 7 billion to 9.3 billion by 2050. Because of increasing urbanisation [21] most of the additional 2 billion plus will impact urban areas in developing countries. In terms of infrastructure development this is compounded by people living in smaller units [22]. The development of thinking on cities of the future ([23], [24], [25]) will be critical to the realisation of universal access in urban areas.

Whatever their limitations, the Millenium Development Goals (MDGs) [26] have been seen as an important driver for progress. Goals were set to halve those without "improved" water and sanitation by 2015. The question being addressed is what should follow post- 2015. At the time of writing a major consultation process was underway [27] but decisions had not then been made. It was clear that future "targets" and monitoring would reflect the "elements" in the UN resolution. The importance of having ongoing international targets and monitoring was recognised in the report [28] by the Independent Expert in 2010. I quote from the report: "The Joint Monitoring Programme should serve as a global discussion platform to facilitate the formulation of possible next-generation global water, sanitation and hygiene targets, corresponding indicators and adequate data collection mechanisms. It would be well positioned to compile experiences on the measurability of additional indicators, commission research into these issues where needed and disseminate experiences. To meet this task, it will require adequate resourcing."

There are no doubts that the final goal should be universal coverage for both water and sanitation, but with uncertainties of population growth it will be difficult to set a firm date for that achievement. Realistically, progress can be driven better by having staged objectives. It is important that any such targets are not just aspirational but practically realisable. Equally interim targets should not be seen as an end point but as a stage on a journey to universal coverage. Monitoring criteria should embrace key measurements related to the elements in the human rights declaration.

AN OPPORTUNITY TO SUCCEED

With rights there are responsibilities. Those responsibilities include the wise use of water and willingness to pay, so that there are sustainable systems for both current and future generations. Public consultation should embrace all aspects, including understanding of timescales and progressive realisation and for citizens to play their part in sharing the responsibility for the service delivery. So many good intentions and reports are left on the shelf to collect dust. So often dogmas or narrow interests get in the way of progress. We must not let this happen on human rights. We must avoid good intentions turning into barriers to progress. Universal access is possible. We must focus on that objective.

REFERENCES

1. Resolution adopted by the UN General Assembly 64/292. The human right to water and sanitation. 3 August 2010.
2. O'Neil O. (2005). The darker side of human rights: a revised version of the *Martin Wight Lecture* given at the London School of Economics on 14 October 2004. *International Affairs*, **81**(2), 427–439.
3. Bakker K. (2010). Privatizing Water: Governance Failure and the Worlds Urban Water Crisis. Cornell University Press. ISBN: 9780801474644.
4. Tortajada C., Joshi Y. and Biswas A. (2013). The Singapore Water Story: Sustainable Development in an Urban City State. Routledge, London.
5. Human Rights and the Environment Rio+20: Joint Report OHCHR and UNEP http://www.unep.org/environmentalgovernance/Portals/8/JointReportOHCHRandUNEPonHumanRightsandtheEnvironment.pdf (accessed May 2013)
6. The Universal Declaration of Human Rights – the United Nations (1948). www.un.org/en/documents/udhr/history.shtml (accessed May 2013)
7. General Comment 15. UN 2002. www.unhchr.ch/tbs/doc.nsf/0/a5458d1d1bbd713fc1256cc400389e94 (accessed May 2013)
8. Onda K., LoBuglio N. and Bartram J. (2012). Global access to safe water: accounting for water quality and the resulting impact on MDG progress. *International Journal of Environment Research and Public Health*, **9**(3), 880–894.
9. Constitution of the Republic of South Africa [No. 108 of 1996]. http://www.info.gov.za/documents/constitution/1996/a108-96.pdf (accessed May 2013)
10. Water Services Act [No. 108 of 1997]. South Africa. www.dwaf.gov.za/Documents/Legislature/a108-97.pdf (accessed May 2013)
11. The Rights to Water and Sanitation in National Law. Information Portal on the Human Rights to Water and Sanitation. http://www.righttowater.info/progress-so-far/national-legislation-on-the-right-to-water/ (accessed May 2013)

12. Report of the Independent Expert on the Issue of Human Rights Obligations Related to Access to Safe Drinking Water and Sanitation, Catarina de Albuquerque. UN A/65/254, www.ohchr.org/Documents/Issues/Water/MDGReportA6524.pdf (accessed 6 August 2012).
13. On the Right Track: Good Practices in Realising the Rights to Water and Sanitation. OHCHR March 2012. www.ohchr.org/Documents/Issues/Water/BookonGood Practices_en.pdf (accessed May 2013)
14. Catarina de Albuquerque. Report of the independent expert on the issue of human rights obligations related to access to safe drinking water and sanitation. A/HRC/12/24, UN General Assembly Human Rights Council. http://www.un.org/waterforlifedecade/pdf/05_2011_human_right_to_water_reader_eng.pdf (accessed 1 July 2012).
15. Rouse M. (2013). Policy brief: the urban water challenge. *International Journal of Water Resources Development*, http://dx.doi.org/10.1080/07900627.2013.791568.
16. Response from Brazil on Human Rights on Water and Sanitation Independent Expert's Questionaire on Sustainability (2013). www.ohchr.org/Documents/Issues/Water/ContributionsSustainability/Brazil.doc (accessed May 2013)
17. Barlow M. Our Right to Water: A People's Guide to Implementing the United Nations' Recognition of the Right to Water and Sanitation. www.canadians.org/water/documents/RTW/righttowater-0611.pdf (accessed May 2013)
18. Doshi V., Schulman G. and Gabaldon D. (2007). 'Lights! Water! Motion! In Global Perspective'. Strategy + Business Issue. Booz, Allen, Hamilton, Washington DC.
19. UNICEF and World Health Organisation (2012). Progress on Drinking Water and Sanitation, 2012 Update, http://www.unicef.org/media/files/JMPreport2012.pdf (accessed May 2013)
20. UN (United Nations) (2011). World Population Prospects: The 2010 Revision. United Nations, New York.
21. UN (United Nations) (2012). World Urbanisation Prospects: The 2011 Revision. United Nations, New York.
22. Onda K., LoBuglio N. and Bartram J. (2012). Global access to safe water: accounting for water quality and the resulting impact on MDG progress. *International Journal of Environment Research and Public Health 2012*, **9**, 880–894; doi:10.3390/ijerph9030880.
23. Cities of the Future: IWA WaterWiki. http://www.iwawaterwiki.org/xwiki/bin/view/Organizations/+Cities+of+The+Future (accessed May 2013)
24. Novotny N. and Brown P. (eds) (2007). Cities of the Future: Towards Integrated Sustainable Water and Landscape Management, IWA Publishing, London. ISBN: 9781843391364
25. Hao X., Novotny N. and Nelson V. (eds) (2010). Water Infrastructure for Sustainable Communities(s), IWA Publishing, London. ISBN: 9781843393283 Pages: 500. Paperback.
26. United Nations (2000). Millennium Declaration 55/2, http://www.un.org/millennium/declaration/ares552e.htm (accessed 8 September 2000).
27. The World We Want: The Water Thematic Consultation Report: UN Water. http://www.worldwewant2015.org/node/341163
28. Human rights obligations related to access to safe drinking water and sanitation UN A/65/254. http://www.un.org/waterforlifedecade/pdf/05_2011_human_right_to_water_reader_eng.pdf (accessed 6 August 2010).

Chapter 2

Regulation and regulatory bodies

Friend or foe?

I see some form of independent regulation as essential for sustainable water services. There are those who believe that integrity in local government is sufficient, and others who believe that "the market" can achieve everything, and they would disagree with my statement. I leave it for them to justify their arguments. This chapter gives my arguments as to why regulation is essential, and why some form of independent regulatory body is necessary in most situations. There is a great deal of scepticism as to whether regulators can be truly independent, and this depends critically on the political climate in a country, and on the will of the people. Thus, the relationship of regulators with public representative bodies is critical. It is important at this stage to stress that "one size doesn't fit all", and the appropriate form of regulatory body depends upon a country's situation or the local circumstances.

Regulation of water services only really became an international issue following the "privatisation" of the Regional Water Authorities in England and Wales in 1989, and the interest this stimulated in the greater use of the private sector elsewhere. There arose a concern over the control of private monopolies and the UK Government used the experience of privatisation of the energy and other sectors to introduce a regulatory regime. A 1997 paper [1] on regulation of privately operated water services said "there is widespread agreement that independent regulators are at the heart of regulatory governance for liberalised sectors", referring to OECD, the World Bank, IMF and Regional Development Banks. Due to the ineffectiveness of some regulatory bodies, more recent papers [2] are now questioning that earlier statement, and I consider the competing arguments later in the chapter. However, in 1997, although there was enthusiasm for regulation of the private sector, there was less or even no recognition of the need for regulation of the public sector. As will be seen in Chapter 7, there is now a growing recognition of an equal need for regulation of publicly managed water services to achieve both transparency and efficiency. The UN Human Development Report 2006 [3] referred to the need for regulation of publicly managed water services, with one of its recommendations for government action being "regulating water utilities to improve efficiency, enhance equity and ensure accountability to the poor". In this chapter, I make no distinction between regulation of public and private operations.

There are four areas of regulation. Three of these can be described as economic regulation, drinking water quality regulation and environment regulation, all of which are covered in this book. The fourth area, relating to the enforcement of general business and commercial law, such as those associated with accounting practice, employment legislation and health and safety at work, is not covered, as it is not peculiar to water services. This chapter covers the broad canvas of regulation and regulators and Chapters 3 and 4 cover the more detailed aspects of planning, finance and cost recovery, and standards, monitoring and enforcement respectively. We need first to have a clear view of what we mean by regulation.

WHAT IS REGULATION?

Regulation has been around for as long as we have had laws that relate to the protection of citizens. Police are monitors and enforcers. In 1985, Selznick [4] described regulation as "sustained and focussed control exercised by a public agency over activities that are valued by a community". Tenenbaum [5] in 1995 referred, in a power sector context, to "regulation (as) a system that allows a government to formalise and institutionalise its commitments to protect consumers and investors". Baldwin and Cave [6] point out that although regulation is seen generally as a means of control of activities to prevent something undesirable from happening, regulation can also be a means of facilitating something good such as efficient and disciplined use of radio frequencies. This is important, not just in the form of regulation as in the Baldwin and Cave example, but in the way in which a regulator operates. Within "a command and control" regulatory regime there is a balance to be made between the use of the carrot and the use of the stick. Not only does the carrot produce quicker results, but it prevents pressures developing for deregulation. Regulation managed in a "sensible" way can be seen as desirable, rather than a burden, to those being regulated. I will return to this later. McIntosh [7], in the context of the development and management of urban water supplies and sanitation in developing countries, sees "regulation (by definition) as being about making and enforcing rules". Jamal Saghir [8] warned that "regulation is not a substitute for governance". I agree with that but, at the beginning of Chapter 1, I gave a list of elements which together I see as constituting good governance, and two of these relate to regulation. One is "regulatory bodies that are independent and have the power to implement policies, laws and rules in a transparent way", and the other is "effective monitoring and enforcement of regulations". I see regulation as the process of interpreting and implementing laws, policies and regulations, to achieve what was intended in their formation. This is a pragmatic definition which can be at odds with the views of lawyers. There is another reason why the carrot is better than the stick. The stick can result in litigation in an enforcement process with undesirable outcomes. The use of courts by a water regulator is essential at times, but I see it as a failure, by a regulator, to persuade those it regulates to meet the requirements, when it has to resort to legal sanctions. The "carrot and stick" argument is complicated by the attitude of the media in assessing the reputation of a regulator according to its prosecution "score". This can be compounded by pressure groups who wish to have their "pound of flesh". So the general definitions of regulation show that it can cover a wide range of different aspects. In the water sector, even though there is more quality than economic regulation, the word regulation can be seen as economic regulation, so it is important now to give a specific definition to that.

WHAT IS ECONOMIC REGULATION?

It is important to distinguish between the wider roles of regulators and the aspect which is economic regulation. The World Bank Explanatory Notes [3] discuss "economic regulation versus regulation generally", and conclude with a core definition of economic regulation as "the rules and organisations that set, enforce and change the allowed tariffs and service standards for service providers". The Notes give a valuable illustration of the overlaps with other aspects of regulation, with the source attributed to Castalia, which I reproduce in Figure 2.1. The core definition covers only the non-overlap portion labelled monopoly, but in practice economic regulators also have varying responsibilities for some of the overlaps. I discuss the advantages and disadvantages of aggregation of these functions in a later section.

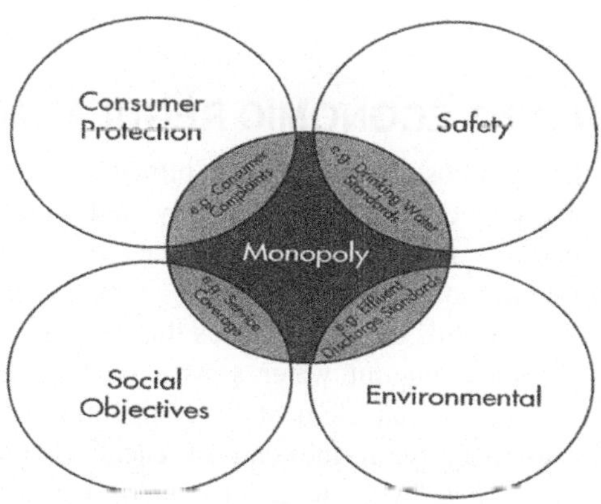

Figure 2.1 Overlap between regulatory functions (*Source*: Castalia).

WHY NOT SELF-REGULATION?

Eijlander [9] gives three characteristics of self-regulation: it concerns the regulation and ordering of behaviour in a certain group of society; it consists of rules which have been laid down for and work within the group; and these rules can be enforced against the group members. In the water sector, forms of self-regulation have been practiced by water service providers for decades through their utility or professional associations, although I doubt that those involved in the early years would have recognised it as self-regulation. Generally this has been limited to technical matters in setting technical standards and codes of practice, although in recent years it has been extended to the establishment and management of benchmarking schemes, as in the Netherlands and Australia. In the Netherlands, the public sector water service provider managers see it as part of a defence against privatisation and associated independent regulation. I am not aware of any countries where water charges are self-regulated by utilities through an association.

Self-regulation can be introduced as part of seeking recognition and respectability. In Ghana, private water tanker operators have formed associations in which members operate to a Code of Practice. This Code has been developed with the regulator (the Public Utilities Regulatory Commission), but is enforced by the association members themselves. This should provide benefits to the communities served by the tankers, providing for discipline on charging and the prospect of better control of water quality. The benefits to the private operators should be that they have become recognised as an important part of the public water supply distribution system, and should be in a better position to negotiate with Ghana Water Company on water charges.

With the benefits of low cost, and with good knowledge of costs and investment requirements within utilities, why shouldn't self-regulation be the norm? After all, it is the form of regulation usually practiced by local authorities. Unfortunately, it has generally failed to produce efficient and sustainable water services, due to water charges being set for political expediency rather than being determined objectively.

WHY INDEPENDENT ECONOMIC REGULATION?

Essentially, due to the high cost of the required infrastructure, particularly water mains and sewerage pipe networks, water supply and sanitation services are monopolies. Some parts of the UK have introduced or are introducing direct competition for non-household customers using common carriage, which is discussed in Chapter 10, but this chapter assumes that there is a need to manage a monopoly situation. Contracting out water service management to the private sector provides for the occasional element of competition, at the time of the initial tender, and subsequently at the end of each concession, but this is infrequent and the choice rests with the asset owners, not with the consumers. Contracting out has introduced the term "regulation by contract". There are differing views as to whether this is regulation and it is discussed later in the chapter. Consumers cannot rely on the market to protect them from higher than necessary charges, from poorer than required quality of service, or from quality of water that does not meet specified standards. So service quality and "product" quality have to be defined, and the price for their delivery has to be determined through some other mechanism.

Generally, water services are provided by local government, with charges fixed internally by those local authority suppliers, either in a recognisable water service bill, or as part of local taxation. This is described as self-regulation, and it is argued that this is an effective method, because if the responsible officials have been elected, they have a duty to serve their constituents and failure to do so will result in their removal at the next election. However, because budget determination and price fixing are internal processes, and only one part of a much bigger local authority portfolio of functions, there is little transparency, so consumers do not have an understanding of the relationship between water charges and the cost of providing the services. As a consequence of this, and with little direct public participation, increases in water service charges under this self-regulatory regime are not only unpopular, but also viewed with suspicion. So, instead of overcharging, local political pressures have tended to result in undercharging, with water service prices set too low for sustainable cost recovery. The situation is compounded in situations where water income is used to subsidise other local services, leaving even less money for maintenance and refurbishment of the infrastructure. There are exceptions to this gloomy picture, with some very good local authorities operating self-regulation in an effective and transparent way. These exceptional examples are described in Chapter 7. Even in these cases, changes in local political leadership could result in a change in policies and priorities with a consequent risk to performance of water services. In Chapter 1, as a major policy consideration, I argued the need for separation between policy, regulation and delivery functions. Therefore, although I do not wholly disregard self-regulation, because there are some situations, such as in Singapore, and in cities like Seattle in the United States, where it has worked well, I believe that generally there is a need for some form of independent and objective regulation. The need for separation of policy and delivery is becoming increasingly recognised [10], and public sector reform most commonly involves the establishment of "corporatised public" utilities. The approach taken to this reform, with successful examples, is given in Chapter 7. This separation creates the opportunity for the employment of good operational managers, with objectives focussed on efficient service delivery. It also provides for separate accounting and greater transparency. However, if price fixing, or price approval, continues to rest alongside the policy function, there remains the risk of political expediency in the determination of charges. The need to separate the regulation function from the policy making body has not yet received universal recognition, but to me it is essential. An independent body is necessary to give objectivity to the process of turning policies into reality, so that unrealistic timescales for the ambitions of politicians become identified through the planning and price determination process. This is an appropriate juncture to consider approaches to setting tariffs.

APPROACHES TO SETTING TARIFFS

The regulator, whom economists [11] consider a welfare maximiser and the utility (which if private is considered a profit maximiser) have important constraints. The utility as a monopoly is required to meet all demand at the regulated prices and is restricted in its level of profits. The regulator whilst protecting consumers has to ensure that the utility can finance its commitments. This is a fine balancing act between being too soft on the utilities resulting in higher tariffs and being too tough which impacts on utilities' ability to raise capital for the required infrastructure investment. There are a number of models which are used. To a certain extent the choice for regulation depends upon the ability of the regulator to obtain good data comparable with that of the utilities. Six models are discussed briefly; iterative, yardstick, rate of return, price-cap, revenue-cap and efficient firm.

An iterative process is useful where a regulator has little information. The regulator sets prices based on the period before. The profit levels and supply volumes for the period then provide the basis for prices for the next period and so on. This process continues until the allowable prices just cover costs based on the previous period. This can be a slow process, and it is difficult for the regulator to take into account the requirements of the poor. Each utility is considered without reference to any other.

In the yardstick model, the regulator uses information on one utility to regulate the others in the group of utilities. The "chosen" utility is then regulated on the basis of information from the other utilities. One argument for this approach is that the utilities do not have an incentive to give false information as their situation is not influenced by their own data. An advantage for the regulator is that this approach requires a low level of information. For this approach to work there has to be a number of utilities with similar demand and cost conditions. To maintain this condition, regulators tend to oppose mergers or takeovers.

With a rate of return approach utilities are allowed to charge prices equivalent to a competitive market. The regulatory aim is to set these charges equal to efficient costs of water services operations plus a market determined rate of return on capital employed. This approach was used almost universally until the 1990s but it is now recognised that it can result in "cost-padding". Also it is criticised as encouraging use of capital rather than labour, although this charge can equally be levelled at some other approaches in which cost of capital is a critical factor in the tariff calculations.

In the price-cap approach the regulator calculates limits (caps) on tariffs using the formula CPI-X (ie Consumer Price Index less an efficiency factor). X is dependent not only on the past performance of the utility, but also on the performance of the other utilities in the group being regulated. The use of the X factor is considered a proxy for the market. The regulator has to avoid using "acceptable rates of return" in the X calculation to prevent it becoming rate of return regulation. Price-caps as used in England and Wales are described in Chapter 10.

Revenue cap is similar to price-cap but with the cap on total revenue rather than prices. Less used than price-cap but useful for sectors (such as electricity) where possibly the utility can have less influence on the demand, and where costs are independent of short-term variations in demand. Both price-cap and revenue-cap approaches provide incentives for companies to outperform, as they can be set up to allow utilities to retain additional profits until the next price review.

Efficient firm regulation is similar to yardstick regulation but uses the concept of the "perfect utility" to identify opportunities for efficiency improvements. It involves setting prices equivalent to long-term average costs. This is optimal when the utility is required to be self-financing. It requires the regulator to have precise information in order to define the perfect utility. Its use in Chile is described in Chapter 9. The perfect utility concept can be used by utilities themselves to set their own performance targets. Dunea, the water utility in The Netherlands serving The Hague, estimates how much cost saving a perfect Dunea could achieve and sets targets accordingly.

One preferred approach is some form of CPI-X formula to determine price caps, but arguments between economists on the relative effectiveness of approaches continue. CPI-X has been described (12) as "rate of return regulation with a lag". Nevertheless the system in England and Wales can be considered to have delivered a major investment programme without excessive increases in charges. To my mind the choice of tariff calculation formula may be less important than a periodic planning system which provides for all requirements to be considered in an integrated way. Associated with this is the important question of how to determine investment programmes to refurbish neglected underground infrastructure. This is discussed in Chapter 3.

WHY INDEPENDENT DRINKING WATER QUALITY REGULATION?

The responsibility for regulating drinking water quality often rests with those local authorities who are also responsible for the water service delivery, so in these circumstances there is self-regulation. This can be very effective, with those responsible for monitoring and reporting drinking water quality being dedicated people with high integrity. Difficulties can arise when there are competing demands for investment to meet standards, or for example if a local politician would rather that a report on a water quality incident was not made public just before an election. So the arguments made above for independence for economic regulation apply equally to drinking water quality regulation. Above all, an independent regulator can ensure that there is transparency in the reporting on drinking water quality and on the investigation of incidents affecting drinking water. It also provides an independent audit function necessary for the attainment and maintenance of high professional standards within water supply operators. Effective drinking water quality regulation is covered in Chapter 4.

ENVIRONMENTAL QUALITY REGULATION

It is accepted generally that the regulation of the environment is a separate and independent role from that of water supply, because the function goes beyond the water supply chain. The roles were combined in England and Wales during the period 1974–1989 (see Chapter 9) in the operation of the Regional Water Authorities. Although seen by some as utopia at the time, the results were not as hoped for, partly due to financial constraints but probably also due to a conflict of objectives when considering water resource needs. A similar conflict of objectives exists when an environment regulator has responsibility for both environment protection and water resources. Clearly, decisions on new water resources, whether involving new reservoirs, new boreholes, or new river abstractions, have to be considered in the context of the impact on the environment, and the decision process has to balance people's needs against ecological judgements. This difficulty can be compounded by environmental pressure groups generally being better organised than the general public, and being in a better position to influence an environment regulator's decisions. To ensure objective decisions there is a need for a means of obtaining equal input from all public viewpoints and a means of independent review. However, this problem is minor compared with the need for regulation to be independent and separate from government, to avoid the risk of decisions being made "behind closed doors" for political expediency.

REGULATION BY CONTRACT

I begin by stating my concern over the limitations of regulation by contract. Clearly a well-defined contract provides the means of controlling the delivery of the desired performance of a water service operator, whether public or private. That is a major step forward and much better than the "traditional" municipal operation in which there are limited or no performance measures and poor financial ring fencing. What it doesn't do is provide any means of regulating the contract owner, nor does it provide for transparency and public participation. On the contrary, it tends to reduce transparency and assumes that governments can look after the consumers' interests. Also, this lack of transparency risks providing the conditions for corruption. Regulating the contract owner might be considered a strange concept, but over the period 2001–10 there was a good example of this in the United Kingdom. Dŵr Cymru, the now "public" licenced "water undertaker" providing water services to Wales, chose to contract out the delivery of its service to private companies. It controlled that delivery through the conditions in the contracts, but Dŵr Cymru itself was subject to regulation by the economic regulator Ofwat. Any penalties for failures to meet regulations fell on Dŵr Cymru, not on the contractors. The contracts can be subject to commercial confidentiality without fear that the consumers will suffer, because the regulatory process ensures transparency in the process of determining tariffs and in delivery of the service commitments. I appreciate that this was a unique situation, but to me it provided a way forward on the regulation of water services under Public Private Partnership situations. It means that should be some independent regulation dealing with the objectives and deliverables around a contract, assuring transparency on performance. This should not interfere with the contract itself, which should be devoted to achieving the outputs at a competitive price. The Dŵr Cymru example is also relevant to the recommendation in the UNHD Report 2006 already mentioned, which advocates some form of regulation "... to ensure accountability to the poor", as taking this forward would involve regulation of local government.

There have been too many contract failures to assume that it can be left to "regulation by contract". In Chapter 8, on the successful use of the private sector, I give some examples of both successes and failures. The reasons for failures can be attributed as much, or more, to the contract owners as to the contractors. What is often missing is an effective planning process which considers objectives, costs and charges in an integrated way. If this process is handled by an independent regulator, unrealistic political ambitions, such as solving infrastructure problems at the same time as cutting overall costs whilst providing additional income for a city, will become apparent. Once realistic deliverables have been established and defined clearly, a contract can be placed by the local government for delivery. The cost of the contract together with other costs for sustainability of the water service can be incorporated into charges, and in subsidy provisions, in a transparent way. There is also the need to have a means of handling changing circumstances, whether they are political, legal or physical, and this requires a periodic review. This needs to take place at least every five years, and is best handled by a regulator. The contract needs to include provision for such reviews, and provide the basis for the contract conditions to be adjusted to reflect the changed delivery requirements, with a corresponding price adjustment. This process should avoid "fudging" (i.e. "glossing" over, or ignoring) issues for political expediency, provide contractors with a stable commercial environment, and also give them respectability. A regulator is in a position to achieve necessary public participation in a way which will focus on people's needs and on the concerns they have for affordability, whilst avoiding the whole focus being on the emotive issues associated with the use of the private sector. So, to my mind, a contract provides for a sound basis for monitoring delivery, whether by a private or public service provider, but it is not sufficient for regulation, which needs to include the process for determining realistic objectives, how costs will be recovered for sustainability, how changing conditions will be accommodated, and the involvement of the public.

The World Bank Explanatory Notes [2] warn about the problems of mixing the two "traditional" designs of regulation, referring to them as the French contract-based tradition and Anglo-American style regulation, and give examples from around the world where hybrids have failed. Considering those cases, the failures appear to have been due to interference by the regulator in the contract details. As I have explained above, I don't see the regulator as interfering with the contracts, but providing the objectivity and transparency on which the contracts can be based. The Notes, in considering how there may be benefit in involvement of a regulator, refer to "creating an autonomous body to monitor the contract, enforce it, and provide the analytical input when tariffs or other aspects of the contract are reset or renegotiated". This is consistent with the process I have outlined in the previous paragraph. Concern should go beyond having efficient delivery in the short term, to having sustainable water services in the longer term, and this means sustainable cost recovery. History tells us that governments, without the benefit of an independent regulator, do not find it possible to set, or maintain, a level of water charges necessary to achieve sustainable cost recovery. However, there is a tendency to increase charges at the time of awarding a concession contract, to pay for the cost of service improvements built into a contract. This leads to public resistance to the use of the private sector, with the risk that pressure groups, opposition parties and the media will look for any opportunity to undermine any ongoing commercial commitments. Independent regulation, with effective public participation, can alleviate such a situation, through providing a climate of better public understanding and trust, which regulation by contract is unable to do.

REGULATORY BODIES

Regulatory bodies vary in scope and size. They can be dedicated regulatory bodies with strong in-house capability, such as those in the UK, or they can be a small part of another organisation without a regulatory-type name and rely on outside professional assistance as and when required. A regulator can deal wholly with water tariff setting as in the core World Bank definition or it can be responsible for all aspects of economic and quality regulation. In some countries, economic regulation of all utilities is handled by one regulatory body. Regulators have different degrees of autonomy with some being wholly independent. My definition of independence is that an individual, or a regulatory body, has the freedom to set tariffs, take enforcement action, impose sanctions, talk to the media, and report on any aspect of the regulatory functions without reference to anyone else. In considering what works well in which circumstances, it is instructive to consider some examples of regulatory systems.

England and Wales

Across the UK, there are strong regulatory bodies for both private and public systems. In England and Wales, there are separate independent bodies covering economic, drinking water quality and environmental regulation, each meeting my criteria for independence. There is also an independent Consumer Council established to give consumers a voice. The functions of these are discussed in Chapter 9. The cost of regulation of the four bodies combined, but excluding the costs the water companies incur in the regulatory process, is less than £1 sterling per consumer per year, some of which is levied in direct water charges, and with the remainder coming from general taxation. The system has delivered improved quality of services, secured drinking water quality virtually 100% compliant with the EU Directive, and provided for the investment necessary to achieve the desired water environment objectives. Importantly, the system has also provided for investment to deal with a backlog of work on the ageing underground pipe systems. Through improvements in efficiency, charges have increased by around half of that which would have been required without a strong regulatory system.

As mentioned earlier, the water company in Wales, Dŵr Cymru, comes under the same regulatory system as the water companies in England, even though it is a form of corporatised public company.

Scotland

The regulatory system in Scotland is similar to that in England and Wales, with four equivalent bodies regulating a single but large public utility, Scottish Water. Until 2005 the economic regulator had an advisory role, but now the Water Industry Commission independently fixes limits on water charges based on output objectives defined by the Scottish Parliament. Scotland has started later than England and Wales to invest in its water and sewerage infrastructure but has a major investment programme to recover its serviceability. Interestingly, Scottish Water formed a company Scottish Solutions, to partner with the private sector to deliver this large construction programme. Partnering is discussed in Chapter 8.

Northern Ireland

In April 2007 water economic regulation became the responsibility of the utility regulator covering electricity, gas and water. Currently the Northern Ireland Authority for Utility Regulation (UREGNI) regulates service standards and costs of Northern Ireland Water, a corporatised public company (a GoCo) similar to Scottish Water. However as 76% of the income (13) comes from government it is also classified as a Non-Departmental Public Body, making it a “hybrid”. The regulator fixes tariffs for industrial and commercial customers. It had been the intention to introduce domestic customer water charging in 2010 but as a result of strong public opposition that has been postponed. Historically, water service charges had been included in local taxation and the public argued that they were already paying for water services by that route. At the time of writing there is a proposal to phase in water charging with a corresponding reduction in local taxation. Major investment in refurbishing a deteriorated infrastructure is required so it is important that the water service in Northern Ireland evolves to sustainable cost recovery.

Chile

Following the Tariff Law of 1988, the economic regulator, Superentendencia de Services Sanitarios (SISS), was formed. SISS works in a similar way to Ofwat in England and Wales, in regulating the private sector, although SISS is also responsible for overseeing concession contracts, for establishing and monitoring compliance with technical standards, and for the control of wastewater discharges. Its role is to protect consumers on charges and to ensure adequate investment in water services, but it is not required to operate as a social service. Tariffs are set to achieve full cost recovery. Subsidies, which are directed to assisting the poor, are handled by Local Authorities. The Chile story is covered in more detail in Chapter 8.

Zambia

The National Water Supply and Sanitation Council (NWASCO), despite operating with very limited resources, is regulating significant improvements in water and sanitation services. The structure of reform, giving autonomy to the regulator and with sound policies, is a model not only for developing countries, but also for the world. Although there had been considerable earlier discussion on the need for reform of the water sector, reform [14] did not start until 1994 with the National Water Policy. This was taken forward by a group of representative stakeholders, supported by the Water Sector Reform Support Unit. One result was the Water and Sanitation Act 1997 which set up NWASCO. NWASCO, with just 15 full-time staff and some part-time inspectors, is responsible for economic and quality regulation of water and wastewater utilities in 9 Commercial Units and 22 Local Authorities. In spite of limited investment in the infrastructures, the regulatory system has been very successful, making significant progress towards full-cost recovery, one of the listed objectives of reform. NWASCO uses ten indicators to measure progress. They are water quality, collection efficiency, metering ratio, hours of supply, cost recovery based on operation and maintenance costs, reduction in non-revenue water, staff efficiency, regulator's perception, increase in sanitation coverage and increase in water service coverage. Comparative competition in service delivery based on attainment of defined benchmarks is used to stimulate and reward performance including awards presented annually at the launch of sector reports.

The Zambia Sector Report 2006 [15] describes what has been achieved and the performance measures which have been applied. The report lists the lessons to be learned, which include strong political will, a clear vision and set of principles to give consistent guidance and, very importantly, the need for a regulator to have autonomy. The report stresses the importance of regulatory regimes being tailored to the specific challenges of a country.

More recently good progress (16) is being made in increasing coverage, see table below.

Table 2.1 Increase in coverage in Urban Zambia from 2010–11 to 2011–12.

Date	2010–11	2011–12
Urban Population	5,559,343	5,620,358
Water Service Coverage %	77.5	81.8
Sanitation Coverage %	54.1	56.7

Source: NWASCO.

However performance on unaccounted-for water (UfW) was reported as unsatisfactory. NWASCO report (16) that overall UfW rose from 45% in 2009/10 to 46% in 2010/11. All of the utilities were below the acceptable benchmark and NWASCO is requiring urgent action for improvement.

Water Watch Groups have been established. These are made up of volunteers from the community that monitor service provision and awareness, and facilitate the resolution of consumer complaints in their respective areas. They are NWASCO's presence on the ground. They assist in improvement in response to consumer complaints, create awareness of consumer rights and obligations and assist in monitoring service providers.

The Government of Zambia has a policy aim of achieving universal coverage of both water and sanitation by 2030. Steps to implement the UN Declaration of water and sanitation as a human right are being taken. See Chapter 1b.

State of Victoria, Australia

In pursuing a reform programme, the State of Victoria consolidated over 400 local authority water service providers into just 20 corporatised public companies. Consideration had been given to privatising the operations, and setting up an independent regulatory system, but due to the successful improved performance achieved with the changes (see Chapter 7), it was decided to retain a corporatised public structure. Later on, the State decided that independent tariff setting would be preferable and brought water regulation under The Essential Services Commission. The Commission covers not only water, but also electricity, gas, railways, ports and grain. The Commission's role is to make price determinations and regulate standards and conditions of service. As in the UK, there are separate agencies for the regulation of drinking water and the water environment. The Drinking Water Regulatory Unit, which has responsibility for the implementation and oversight of the Safe Drinking Water Act, is part of the State's Department of Human Services. The water environment regulator is the State's Environmental Protection Agency.

Manila, The Philippines

Manila is supplied through two concession contracts which, on behalf of the Metropolitan Water and Sewerage System (MWSS) organisation, are regulated by the "Regulatory Office". The Regulatory Office does not set tariffs, which are established in the contract process, but monitors tariffs and service delivery in accordance with the concession contract conditions. The responsibility for drinking water quality and the water environment rest respectively with the Ministry of Health and the Department of Environment and Natural Resources. Manila is one of the examples given in the World Bank Notes [2] in which combining a regulator with a concession contract is considered to have caused confusion. The concession contracts have been controversial with claims [17] of corruption and lack of independence by The Regulatory Office. The Office itself claims independence [18] using independent observers to monitor the concessions. The observers' reports are made public.

The Government (19) has proposed regulation of all water and sanitation service providers whether private or public, including water districts, utilities owned and/or operated by local government units, cooperatives, private entities and those in special economic zones and freeport zones. This was to have covered all operations including water supply, bulk water, sewerage and/or septage treatment and disposal services for domestic/residential, industrial or commercial use. A new regulatory body, to be called the Water Regulatory Commission would have been formed and would have taken over all the current economic regulatory bodies including MWSS and those in local government. However, the 2012 Bill entitled "An Act Rationalizing the Economic Regulation of Water Utilities, Creating the Water Regulatory Commission and Other Purposes" did not obtain the consent of Congress. It is likely that another attempt to achieve a comprehensive regulatory system will be made in the future.

Ghana

In Ghana, economic regulation and drinking water quality regulation are the responsibility of the Public Utilities Regulatory Commission, PURC, which also covers electricity. Water environment regulation is the responsibility of the Environment Protection Agency. PURC regulates one corporatised public company, Ghana Water Company Limited, GWCL, which is the licensed service provider for all urban areas in Ghana. During the period 2006–11 the management of operations was contracted out to an international public sector consortium Aqua Vitens Rand Limited (AVRL) although GWCL continued to be the regulated body. As a distant observer I consider that AVRL introduced good management practice but was limited in what it could achieve by the way the contract was managed. For example operational "capital" for meters and other equipment to improve the operation of distribution systems was controlled by GWCL, not AVRL, and it was difficult for AVRL to have these monies released in order to make progress. I mention this in the context of the importance of effective procurement of private services which I discuss in Chapter 8. Another difficulty for AVRL was that it was working against a background in which propaganda that this contract was privatisation was being used by those opposed to the contract.

PURC recommends tariffs, and has a monitoring role, but decisions on implementation of tariffs are taken by the Government. When the Government does not allow the recommended tariff increases it is intended that it makes up the difference through subsidies. The water distribution systems are in poor condition requiring a major and sustainable investment programme. Average non-revenue water for the urban sector in 2008 was reported (20) at 50.1%. Water is rationed through intermittent supplies with consequential impact on water demand and leakage (see discussion on continuous supplies in Chapter 1 with reference to the experience in India). Although the regulatory structure is good in concept it appears to me that PURC is constrained by lack of powers in practice to obtain information and to take enforcement action.

PURC, as part of its role, is pursuing a pro-poor policy in recognition of the need to extend the service to the poor and unserved. There is an initiative by the Ministry, PURC, GWCL and WaterAid, an NGO, with World Bank funding. It involves setting up water boards made up of local community leaders to manage local systems. Following successful pilots in the capital Accra the approach is being extended to other parts (see public participation in Chapter 5).

Portugal

Instituto Regulador de Áquas e Resíduos – IRAR, the Portuguese Institute for the Regulation of Water and Waste, was established in 1998 [21]. It had limited powers regulating only those operators with concession contracts. However it did regulate drinking water quality across the country. In spite of the limited remit it did make good progress through pursuing a cooperative approach in implementing a benchmarking system, with 20 performance indicators on water supply and 20 performance indicators on wastewater.

In 2011 (based on Decree-Laws 2009) IRAR became ERSAR with responsibility for regulating the whole of the water, wastewater and municipal waste sectors in Portugal (22). This involves 498 operators as shown in Table 2.2

Table 2.2 Numbers of operators regulated by ERSAR.

Management models	Drinking water	Wastewater	Municipal waste management	Total
Direct management	213	217	241	284
Delegation	140	24	23	152
Concession	40	40	13	62
Total	**393**	**281**	**277**	**498**

Source: ERSAR, Portugal.

For each sector the regulator has the two objectives of protecting consumers and maintaining viable operators. The regulatory approaches to achieve these objectives are slightly different for each group of operators. The approach for the state-owned water sector is described as promotion of efficient and affordable tariffs, promotion of the economic and financial sustainability of operators and direct price invention through rate of return regulation. ERSAR is facing a huge challenge in taking forward regulation on this scale but the regulatory structure and powers being introduced, together with the experience of the success of IRAR and the experience gained since 1998, provide optimism for success.

LESSONS LEARNED FROM THE EXAMPLES

The examples cover full equity privatisation, as in England, concessions as in Chile and Manila, corporatised publics as in Scotland, Northern Ireland, and the State of Victoria, a mixed concession and local authority structure as in Zambia, Wales, a private company without shareholders, and Portugal a mixture of concessions, corporatised republics and local authorities.

The successful cases all have the following:

- An independent regulator with the power to both set and monitor tariffs, or the authority to set the service conditions on which concession contracts are based, and the authority to oversee the periodic review of contracts through defining the changed deliverables.
- The management of a periodic planning process so that all aspects of a forward delivery programme, to meet government objectives, can be considered in an integrated way at the time of tariff setting.
- The ability or the obligation to establish public participation.

- Effective regulation is affordable and could be set up in all developing countries. It is possible to achieve a full-system of independent economic and quality regulation at low cost, as demonstrated by NWASCO in Zambia, but this requires political will and dedicated management.

EXPERT PANELS

There are good reasons for the introduction of expert panels. One, as in Manila, is that external observers can strengthen independence and give credibility and confidence in the regulatory process. Another is that often a regulatory body does not have, or cannot afford to employ, the necessary expertise. The argument for the latter is that the expertise may not be available in a country, or it may not be needed all of the time, so it is more effective to bring in experts as required. Such experts can bring the benefit of previous experience thus avoiding some of the mistakes made elsewhere. There are also arguments against the use of external expertise, one being the likelihood that governments may wish to appoint pliant individuals. The pros and cons were played out in a World Bank web-based discussion forum [23].

There are many considerations in setting up an Expert Panel. Firstly should its remit be advisory, or should it have the authority to take decisions, or should it be an appeals board as in Chile? Should it work continuously to give benefit of continuity, or should it operate only at specific times to reduce cost? What should be the balance between locals and those who can provide an international input? If it is advisory, what should its relationship be with the regulatory body if that is some form of committee? If it has the authority to take decisions, is there the risk that it becomes the regulator under a different name? Is there an independent body with public credibility who could appoint the members of a panel?

Contributors to the Discussion Forum referred to alternatives to Expert Panels, one being the establishment of regional regulator associations to share experience. This is happening in Europe, South East Asia, in Africa, and in the Americas, and international networks have been set up by the International Water Association and by WHO. Another suggestion was the use of NGOs. This has been very successful in developing countries in achieving sustainability, particularly where there are many small communities. In Chapter 5, I give some examples of where NGOs have provided the expertise both on the physical construction of water supply systems and on their financial stability.

REGULATION IN ASSISTING THE POOR AND UNSERVED

What are the objectives, and how can regulation help? The requirement is to achieve a situation in which available water is shared more equitably to meet the basic needs of everyone, and to achieve a combined top down/bottom up approach to deliver dignified sanitation to all. The UNDP Human Resources Report 2006 (3) has now been largely forgotten due to the UN Declaration of water and sanitation as a human right becoming the main driver for change. However, it was a valuable report and proposed a number of actions by governments with implications for regulation:

- making access to water a human right with implementation based on 20 litres a day of "clean" water for everyone;
- introducing "lifeline tariffs", cross-subsidies and investment in standpipes;
- regulating water utilities to improve efficiency, enhance equity and ensure accountability to the poor;
- enacting legislation that requires female representation on water committees and other bodies;
- giving priority to the needs of the poor in public investment and service provision strategies for water and sanitation;
- empowering independent regulators to hold service providers to account for delivering efficient and affordable services to the poor.

For these proposals to become reality, firstly there is a need for governments to introduce policies and, where necessary, legislation. Secondly, it requires regulation to be seen in a wider remit than "enforcing tariffs and service standards". It comes much more into Selznick's definition [4] as "sustained and focussed control exercised by a public agency over activities that are valued by a community". Thirdly, independent regulators hardly exist at all in public sector situations in developing countries, and there is a need to establish a means by which local authority providers can be held to account.

As stressed throughout this book, one of the reasons why communities are not receiving a continuous safe supply of water is that there is insufficient cost recovery, resulting in inadequate maintenance of systems, high leakage and thus loss of both water and pressure. As a result the poor are paying vendors far more for their supplies than those who have a reliable piped supply. What the poor cannot do is save money, and thus are unable to pay a lump sum for access, or to pay a monthly, let alone a quarterly, bill. The difficult question is how to break out of unsustainable systems. Experience from around the world, some of which is referred to in Chapters 5, 7, 8 and 9, suggests that ideally, for urban supplies, it requires the following:

- national governments to develop sound objectives and policies, including the need for adequate cost recovery for sustainability:
- independent regulators to implement the policies and monitor compliance by the service providers, which in most cases will be local authorities. The monitoring will need to go beyond tariffs and service standards, and include compliance with system extension and increased access targets, and include effective public consultation and information;
- external financial support to pay for the start-up or refurbishment costs. In particular this should include assistance in extending and reinforcing the distribution systems;

- subsidies directed at the poor to provide assistance on access charges, and either low tariff pre-payment meters or means tested direct support on water bill payments. A good example is the pro-poor initiative in Zambia, which is organised by the regulator NWASCO [14]. Water Utilities are paid to extend services in peri-urban areas, financed by the Devolution Trust Fund;
- local water committees to provide the focus for public participation.

In the absence of the establishment of an independent regulator, such as the one operating in Zambia, monitoring local government compliance with national government policy will have to fall on the Local Water Committees, who will need the assistance of competent NGOs. The Citizens Action initiative (24) by the International NGO WaterAid working with local NGOs could become a major force in holding local authorities to account on service delivery to the poor. On a detailed point, there is the likelihood that in many cases, in the shorter term, individual house connections may not be affordable, and that many people will need to be supplied by standpipes made available within an acceptable distance, for instance 100 metres from their homes. I give an example in Chapter 5, Kolhapur in India, where those with a house connection pay an additional amount of money to the community managed system, so that those having to collect water from standpipes have a free supply. In that case the community is both the owner and regulator, but critically there is total transparency thus giving the consumers confidence in the administration resulting in willingness to subsidise the poor.

CONCLUSION ON THE KEY ELEMENTS OF REGULATION

In all situations, large or small, urban or rural, developed or developing countries, public or private service providers, there are a few key regulatory elements for sustainability of water services. They are:

- Some form of body whose prime responsibility or motive is to regulate water services in the interests of all of the community. This can be a formal independent regulator, an NGO or a Community Group. This list should include democratically elected local authorities, but history tells us that in many cases they cannot be relied on to operate either transparently or in the interest of sustainability. Self-regulation is difficult in circumstances where those in a position of authority have other and often conflicting objectives.
- There has to be full-cost recovery of at least operating and maintenance costs, and this does not generally happen without charges being set by an independent body free from political pressures.

- There has to be some form of representative body for consumers without which there is a lack of understanding of the issues, and a consequent lack of willingness to pay the necessary price for water services. The track record of community-managed systems demonstrates that good public understanding is an important ingredient for success.
- There has to be recognition of the needs of the poor, particularly their inability to save, and thus the need for means of easy payment, or targeted subsidies.

REFERENCES

1. Regulation of the Private Provision of Public-related Water Services. Economic Commission for Latin America and the Caribbean, 1997.
2. Groom E., Halpern J. and Ehrhardt D. (2006). Explanatory Notes on Key Topics in the Regulation of Water and Sanitation Services. Water Supply and Sanitation Sector Board Discussion Paper Series, Paper No. 6, World Bank, June 2006.
3. UNDP Human Development Report (2006). Beyond Scarcity: Power, Poverty and the Global Water Crisis. ISBN: 9780230500587.
4. Selznick P. (1985). Focussing organisation research on regulation. In: *Regulation Policy and Social Sciences*, R. Noll (ed.), University of California Press, Berkeley, California.
5. Tenenbaum B. (1995). The Real World of Power Sector Regulation. Viewpoint Note No 50. The World Bank.
6. Baldwin R. and Cave M. (1999). Understanding Regulation: Theory and Practice. Oxford University Press, Oxford, UK.
7. McIntosh A. C. (2003). Water Supplies: Reaching the Urban Poor. Asian Development Bank. August 2003. ISBN: 9789715613804.
8. Saghir J. (2006). Keynote Address at International Water Association Congress, Beijing.
9. Eijlander P. (2005). Possibilities and constraints in the use of self-regulation and coregulation in legislative policy. *Electronic Journal of Comparative Law*, **9**(1), http://www.ejcl.org/91/abs91-1.html (accessed July 2013)
10. Ontario Water Strategy Expert Panel (2005). Watertight: The Case for Change in Ontario's Water and Wastewater Sector. Report of the Water Strategy Expert Panel. Queen's Printer for Ontario, 2005. ISBN: 9780779479283.
11. Personal Communication.
12. Helm D. (2004). The New Regulatory Agenda. Published by the Social Market Foundation, London.
13. Price Cap Regulation in the UK Water Industry (2012). Research and Information Service Research Paper NIAR 790-12. Northern Ireland Assembly, November 2012.
14. Water Sector Reform in Zambia. NWASCO (2003). http://www.nwasco.org.zm/media/water_sector_reform_zambia.pdf (accessed July 2013)
15. Urban and Peri-urban Water and Sanitation Sector Report 2005/6 NWASCO 2006.
16. Urban and Peri-urban Water and Sanitation Sector Report 2011-2012 NWASCO 2012.
17. Montemayor C. A. (2003). Brief History of Water Privatization in Metro Manilla. The Philippine Water Vigilance Network, Presentation for the Summit for Another World Evian, 2003.
18. Personal Communication.

19. Personal Communication.
20. Annual Report (2008). Public Utilities Regulatory Commission (PURC), Ghana.
21. Baptista J., Pássaro D. and dos Santos R. (2004). The Regulation of the Water and Waste Sectors in Portugal. Instituto Regulador de Áquas e Resíduos, Lisbon.
22. The Role of the Portuguese Water and Waste Services Regulation Authority ERSAR, Portugal 2011.
23. Expert Panels (2005). A Cornerstone for Public Utility Regulation? World Bank, website July 2005.
24. Bridging the Gap (2006). Citizens Action for Accountability in Water and Sanitation. Water Aid, London, March 2006.

Chapter 3

Planning, financing and cost recovery

Water: not a free for all

It is argued that water is a human right and as such, should be free. At the Kyoto World Water Forum in 2003, protesters campaigning against the involvement of the private sector in the provision of water services carried placards to this effect. They disrupted serious discussions on how to provide water services to the currently underserved – discussions aimed at meeting the water and sanitation millennium development goals. The Minister responsible for water in Swaziland put this argument in perspective, when he was reported as saying words to the effect that "certainly water is free, providing you wish to take your bucket up the mountain to the source to collect your water, but if you wish me to make provision for the collection, treatment and delivery of safe drinking water to your home, then I have to charge you for the service". This chapter discusses finance measures for such a service, how necessary costs have to be recovered, and the critical planning process to pull it all together. The United Nations has declared water (and sanitation) as human rights but that declaration does not relate to free water services. See Chapter 1 Part 2 for a discussion on the meaning and importance of the UN declaration.

FINANCE NEEDS AND COST ELEMENTS

The costs of water services can be subdivided into start-up costs, operating costs and provision for renewal of the infrastructure. In many parts of the developing world, investment priorities tend to be focussed on start-up costs in the provision of first-time piped water supplies whereas, in the developed world, the investment requirement has moved to dealing with deteriorated infrastructures. Almost everywhere, there have been periods when there has been inadequate provision for infrastructure renewal, resulting in systems in poor condition with adverse consequences of high levels of leakage, poor water quality and frequent disruptions to service delivery. It is estimated [1] that required investment worldwide over the period 2005–2030 is US$ 23 trillion, greater than that required for energy and transport combined. Much of this is required for refurbishing existing infrastructure due to a backlog of investment arising from inadequate cost recovery. The estimated breakdown by regions is given in Table 3.1. This situation has led to attention being given to the need for full-cost recovery, or sustainable cost recovery, or cost recovery for water service sustainability. In this chapter, I will use the term sustainable cost recovery, the definition of which is given in the IWA International Reference Paper [2] as "costs that are recovered so that a water services undertaking can achieve and maintain a specified standard of service, both for the present and future generations". I begin by considering some important principles on what is to be avoided.

Table 3.1 Estimated investment requirement in water and sanitation.

Estimated Investment Requirement in Water and Sanitation 2005–2030	
Area	**US $ 000,000,000,000**
Middle East	0.22
Africa	0.22
US/Canada	3.62
South America /Latin America	4.97
Europe	4.53
Asia/Oceania	9.04
TOTAL	22.60

Source: Booz Allen Hamilton [1].

WHERE NOT TO BE

Downward spirals, or what are often called "vicious circles", should be avoided. Both terms describe the situations well, although unfortunately the terms are used widely in other contexts. I refer to three circles, the consumer circle, the salary circle and the subsidy circle. I have taken them from a presentation [3] by Eero Kontula, Finnida (see Chapter 7). I don't know who the originator was, but he or she should take a bow to the accompaniment of a fanfare of trumpets. The circles (Figure 3.1) are best considered together as the problems they describe are interconnected, with the fundamental cause being unsustainable cost recovery arising from low tariffs. Low tariffs without subsidies mean poor service, discontented consumers and even lower income. Low income means poor salaries, low motivation and the likelihood of corruption, poor performance, discontented consumers and even lower income. Subsidies mean state control, low initiative, low motivation, poor service and so on.

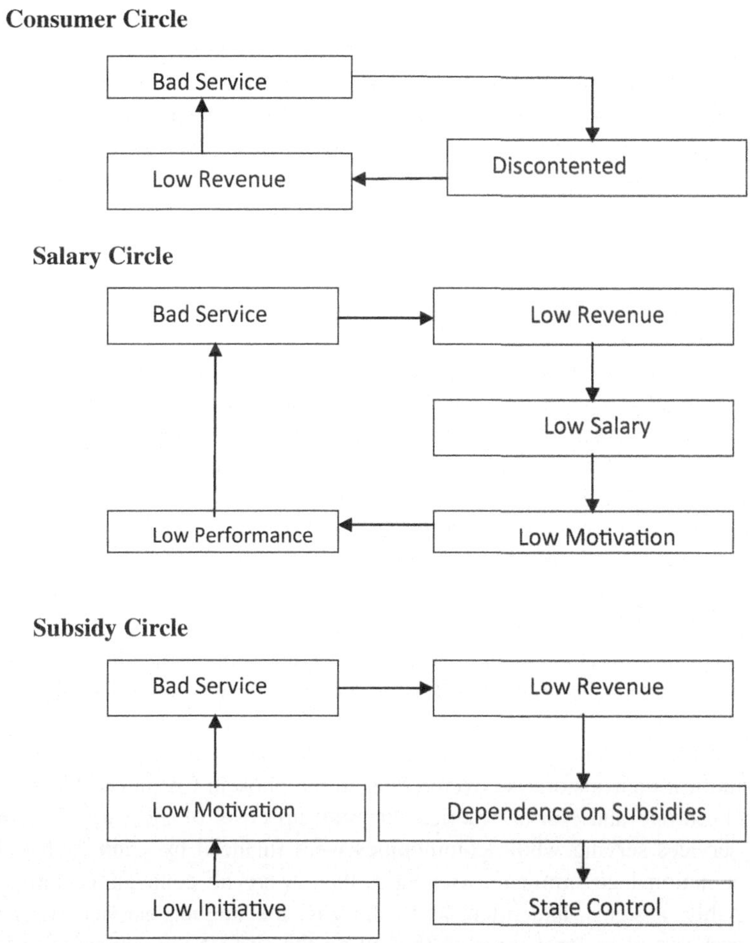

Figure 3.1 The vicious circles of water service decline (*Source*: Presentation by Eero Kontula, Finnida, at World Bank/WaterAid Workshop, London, August 2004; see Chapter 7).

This situation is not limited to the developing world. In parts of the developed world traditional municipal operations suffer from the same ailments. Probably, water sector workers in developing countries would not see their counterparts in the developed world as being poorly paid, but local government pay is often low compared with the private sector. Consequently, local governments find it difficult to attract high quality management and staff to work in the water sector. Elected officials, for popularity reasons, are unwilling to raise tariffs, and subsidies are rarely sufficient to achieve sustainable cost recovery. The consequence is a downward spiral of deteriorating infrastructure and poor service. The problems of the traditional municipal operation, and some solutions, are described with examples in Chapter 7. Some municipalities have looked to the private sector for salvation but without recognising that the fundamentals of investment in the deteriorating infrastructure, and tariff increases for sustainability, are essential. This is quite independent of whether the systems are managed by the public sector or the private sector. Successful use of the private sector is discussed in Chapter 8.

These vicious circles are of even greater consequence in the developing world given the difficulty of raising sufficient funding, and of developing management capacity, as it is more difficult to break out of the downward spirals. There are no quick fixes, it takes time, and there is a need for courageous politicians, either supported or provoked by competent NGOs, to achieve the necessary changes.

I have highlighted where not to be but recognise that, as a result of a lack of attention to important fundamentals, many water service operations are already where they should not be. I now return to the elements of cost. I begin with start-up costs, with the recognition that there are often significant difficulties in building their cost recovery into charges.

START-UP COSTS

In the early days in the development of piped water supplies in Europe and North America, it was entrepreneurial private companies who initially invested in systems for those able to pay for a service. In the UK, these private companies expanded their systems to form statutory private water companies serving whole communities; their history is covered later in the book in Chapter 9. However, in most of Europe and North America, the start-up costs of the early systems of water services serving whole communities were financed by grants, either from local or national governments. Generally there were no connection charges for these public systems, and often, as in the UK, the ongoing service costs were recovered as part of local taxes. Although on the surface this appeared to be an attractive way of establishing services, generally the public were not consulted, and there was no transparency on costs, utility performance or water quality. Consequently, with pressures on local authority finances, it was rare for adequate attention to be given to maintenance and renewal of the system. This was not a problem in the early years as the systems were new, but became increasingly important over time. However, the local community approach had the great value of establishing services for all inhabitants in an area, with a significant beneficial impact on public health.

In some parts of the developing world, the history has been similar, but often the initial infrastructure was installed in affluent areas, leaving a growing and relatively poor population on the extremities of the systems. The challenge now is how to extend systems in a way that is affordable and sustainable. Although the poor, if provided with suitable payment methods, are able to pay for piped water supplies, some people find it difficult to pay for access charges. There are cases where as part of a privatisation scheme, high access charges have been introduced without consultation with consumers or consumer groups, resulting in a low uptake and discontent within the population. One solution, as practiced in Ho Chi Minh city [4], is for there to be no charge for access, with the costs being spread across the water charges of existing users. Another approach is for there to be easy payment schemes, such as in Chile, where connection charges are paid for in easy instalments over five years. This raises the question of the best use of subsidies or grants, which is discussed later.

A very important factor is for the infrastructure to be designed to be affordable. This requires effective consultation with and participation of those consumers who will be affected. If those affected by the installation of additional infrastructure, including both treatment plant and extensions to the distribution system, are brought into the discussion on what is required and what it will cost, there will be the necessary understanding of the changes in charges, and most probably valuable input on payment schemes and tariff structures. Most importantly it is likely to avoid the temptation by authorities to build big or expensively without consideration of the impact on consumers.

An associated consideration is for those who will have to operate systems to be involved in decisions on their concept and design. They need to feel that it is appropriate for their conditions, that they will be able to obtain replacement parts, and that they will be able to operate the systems effectively and efficiently. In some countries, for example Egypt, an organisation has been set up to design and construct new capital installations. The benefit of centralising this function is of having experienced staff able to produce consistent designs, with the ability to manage construction contracts. This should result in economies in both design and construction costs. However, the operating companies should be involved throughout the process to add operational experience, to be involved in the commissioning process, and to feel ownership of the projects right from the outset. Personally, I think it is better for an operating company to have responsibility for design, construction and operations. The benefits of design and construction management experience can be obtained through involving consultants and contractors. Partnering (see Chapter 8) in the UK is proving to be a most effective way of achieving a win-win situation, but with the operating companies in overall control throughout.

So have I wandered away from a consideration of start-up costs? Not at all, as all of the above factors affect both the actual start-up costs and how they will influence the ongoing operational costs and sustainability. This leads on to the important role of donors and international banks in providing funds for start-up costs and what they can do to obtain the greatest impact.

DONOR AND BANK INVESTMENT

The first consideration is where is investment most needed? There appears to be a tendency for funders to wish for a recipient country to invest in new treatment plant or enhanced treatment facilities. From a donor's or banks' point of view, it can be well-defined, costed accurately, and it can be delivered to a predictable timetable. Also there remains a "monument" to an investor's contribution. Local politicians are attracted to these investments in a similar way, especially if the completion can coincide with their period in office, and there is the opportunity of some good publicity. In many cases, additional treatment capacity is the priority, but in many other cases, the priority should be the refurbishment and extension of distribution systems, or a combination of both treatment and distribution capacity. One argument put forward for giving priority to treatment is that long lead times are involved, although in practice the refurbishment of distribution systems takes longer, and is essential in order to deliver increased production capacity.

Due to the downward spiral mentioned earlier, existing distribution systems are often in poor condition with unaccounted-for water as high as 70%, and leakage a major contributor. Supposing leakage is 50% of water leaving treatment works, reducing it to 30% makes 40% more water available for distribution and sale. This is a valuable resource, but reducing leakage also provides for maintaining pressure in distribution systems with the associated improvement in service. Also it greatly reduces the likelihood of contamination of water supplies from ingress of contaminated street water. As discussed in Chapter 1, it is an essential aspect for moving to continuous supplies, so-called 24/7, where as demonstrated in India, less water is required and where consumers are prepared to pay for having continuous availability of safe water.

Also, cities grow not in the best interests of water supply planning, but for many other reasons, including the establishment of unplanned settlements on the outskirts with ongoing movement of people from rural to urban areas. Original distribution systems were designed to supply a defined area with pipes becoming smaller as the branches of the tree network reach the extremities. Therefore, in order to extend systems it is necessary to include investment in larger pipes, either to reinforce systems in the same overall direction away from treatment works, or to create a feed-in from the new extremity back towards the treatment works. The investment required for the latter, and the timetable for construction can exceed that needed for a new treatment works. These considerations require integrated planning, which requires sound network analysis, which in turn requires good information on the distribution systems, and sound water demand predictions. Well, the reader may say, this is all very obvious, but many decisions appear to be made without a good understanding of the options and the knock-on implications of any one aspect. The strong recommendation is that investors should begin by providing support for good technical assistance, to assist the operational utilities in learning how to carry out measurements required for accurate water balances [5], and for calibration of water network analysis models. These tools then provide a firm basis for forward planning on leakage control, extensions to distribution networks and increased resources and treatment. The message here is to start with consumers' needs and work back up the supply system. An excellent example [3] of this is the Finnish Government's support to Vietnam described in Chapter 7. Perhaps in this project a way could be found of sticking a prestigious plaque on a distribution system. I suggest a monument is erected in the main city square which says "Beneath these streets is a refurbished and revitalised water distribution system created by the staff of your water supply company, with technical assistance provided by the Government of Finland, which is providing you with safe drinking water 24 hours a day". This may sound flippant, but something like this is needed to give funding and work on the distribution systems the necessary status and recognition.

The investment needs will be compounded by population growth, estimated [6] to be a further 2.3 billion people by 2050, an increase of 33%, largely in developing countries. With ongoing urbanisation, this means an estimated additional 2.6 billion urban residents by 2050 [7]. The challenge is to expand water and sanitation provision in urban areas in developing countries by at least 50% over and above the current deficit in just 37 years. The integration of water and sanitation planning into sustainable city planning will become more and more critical. Planning is discussed later in this chapter.

A vitally important consideration in the case of loans is their cost and payback period, due to the impact on water charges. Water by unit volume is a low cost product. Capital assets are expensive, and can have a big impact on the cost of water supply, unless the loans have a low interest rate and recovery can be over a period reflecting the life of the asset, say up to 30 years or more. It is for this reason that some concession contracts are for a long period, with cities foregoing the periodic opportunity of competition through re-tendering. Malaysia, recognising the need for long-term low-interest loans, as part of the water sector reforms enacted by the National Water Services Commission Act and Water Services Industry Act 2006 (WIS Act), has formed a government-owned body called Water Asset Management Company (Pengurusan Ast Air Berhad, PAAB) [8] to fund and develop the water infrastructure. This company, backed by government security, is able to secure long-term low-interest loans from the international money market. The approach is for PAAB to fund developments approved by the regulator and to recover the costs through leasing charges on the operator. At the end of the leasing period ownership will transfer to the operator and that capital will become available for other developments.

WATER CHARGES

For sustainability, water charges should be at a level sufficient to cover fully the costs of operating efficient water services, and to make provision for the necessary refurbishment of the infrastructure. As discussed in the last section, the costs to be recovered may include an element for servicing the capital cost of the infrastructure, depending on whether or not there are grants, subsidies or loans. Determining accurate operating and maintenance costs, and calculating production and distribution unit costs, should be a straightforward process, but often in municipal systems, there have not been separate accounting systems, there is a lack of a cost-management culture, and there is no accurate information on water produced or water delivered to consumers. The need for a good water balance has already been mentioned, but there is also a need to identify actual costs of labour, energy and bought-in items. This then provides a sound basis for considering what the costs should be in a reasonably well-operated system, because consumers should not have to pay for poor procurement practice or for the cost of an inefficient service. Achieving a highly efficient service takes time and is covered in Chapter 6, but there can be some good initial gains from good housekeeping.

Often, at the beginning of a reform process, charges are not sufficient to cover even the day-to-day operating costs. Sometimes, there are government subsidies to cover part of the operating costs. However, this is poor practice for three reasons. Firstly, subsidies rarely make sufficient provision for infrastructure refurbishment. Secondly, they are unreliable because there are always competing local needs for those monies, with the inevitable consequence that there will be cuts in the amount available for the water service. Thirdly, general subsidies provide greater benefit to the more wealthy consumers, who use more water, than to the poor. Subsidies, where available, should be focussed on the needs of the poor, and this is discussed later in this chapter.

Given good housekeeping, and a plan to improve operational performance over time, there is a need gradually to increase tariffs towards sustainable cost recovery. Consumers are going to be unhappy if tariffs are raised without some tangible improvement in the service, perhaps seen as more hours of water supply per day following some initial effective leakage control. Consumer consultation is particularly important during the transition stage, with explanation of why sustainable cost recovery is essential, and if necessary a plea for some tolerance during the initial stages, as improvements in service may not be seen in the early stages. Indeed, it may be necessary to advise consumers that for a period they will be without a piped supply, and that a tanker service will be provided, while critical parts of the distribution system are being replaced or refurbished. It is important that consumers are given an honest account of the difficulties, and that expectations are not raised above what can be delivered, especially in relation to the time it takes to achieve improvements. It is easy to allow systems to deteriorate, and loss of a good quality service is not immediate, but once systems are in a state of decay, recovery takes time, investment and staff development. How can financial provision be made to prevent deterioration?

ADEQUATE INVESTMENT IN THE EXISTING INFRASTRUCTURE

Maintaining an existing infrastructure requires good information on the performance of systems, whether they are above ground, as with water treatment plant, pumping stations and treated water storage facilities (easier), or underground, such as pipes and valves (more difficult). The difficulty is that systems don't always give early warning of impending failure, and at the point of failure, the systems, especially those underground, are then in an advanced state of decay. Because of this, it has been easy for those responsible for releasing monies for investment to say, "replacing those items can wait for another year", but that becomes another year and another and another. Equally, it is not an economic proposition to have a policy of replacing all pipes say every 25 years, because age is only one factor, as there is high variability in the life of pipes depending upon a number of factors, including soil conditions and the type of water being distributed. However, it is clearly obvious that an average replacement/refurbishment rate of several hundred years is inadequate. Asset management planning has become one of the most important subjects in water management, beginning with the pioneering work of the UK Water Research Centre (WRc) starting in the late 1970s/early 1980s. Manuals for both sewerage system and water distribution system rehabilitation were produced in 1984 and 1989 respectively. The 4th edition of the Sewerage Rehabilitation Manual was published in 2001 [9]. Since that time there has been increased attention around the world to the subject of asset management ([10], [11] and [12]).

So the most appropriate way ahead is to have good predictive tools for system deterioration and for replacement/refurbishment to take place ahead of the point where the service quality is affected and ahead of the point where action becomes unnecessarily expensive. The costs of doing this are therefore those predicted to maintain the required quality of service on a continuous basis. During the period leading up to privatisation in England and Wales and the production of the first asset management plans, this was referred to as renewals accounting. Unfortunately, standard depreciation accounting systems, used almost universally, are not effective for assets with lives of more than about 10 years, and assume that the value of those assets will diminish over time. Whereas this is true of say a vehicle or a computer, the value of a pipe is the hole it creates in the ground. The value of that hole does not diminish with time, and the important consideration is the cost of maintaining that hole, or rather maintaining its capacity to transmit water without an adverse change in water quality. Therefore standard accounting systems do not make adequate provision for the replacement of those assets.

In England and Wales, infrastructure refurbishment costs are not treated as capital, but are included in operational budgets as an infrastructure renewals charge. These costs are included in the profit and loss account. This is fine providing that regulators take refurbishment seriously, and there is an effective method by which water companies can justify the ongoing annual costs. The risk, as with local authority budgets in which there is a tendency "to leave refurbishment to next year", is that a regulator may be tempted to squeeze refurbishment budgets in favour of lower short-term water charges. An effective depreciation accounting system would bring greater transparency to any shortfall in investment. There is a need for accountants to devise a depreciation system that reflects the long life of assets, so that automatic provision is made from the time a new system is installed. In the meantime, operators need to work hard at making a case for including ongoing asset replacement as part of operating costs, as in the UK.

MAKING PROVISION FOR THE POOR

I mentioned the question of affordability in Chapter 1, with the poor in the developing world paying water vendors anything up to 25 times [13] more for a litre of water than those who have a subsidised tap water supply. The poor do afford water, they have to for survival, but what they would like is a reliable safe supply at a predictable price. What the poor cannot do is save money to pay for an access charge or to pay monthly bills. They need a means of paying an access charge in easy instalments and of paying for water "as they go". Most commonly, there are general subsidies which give most benefit to the "rich", but which are generally insufficient for sustainability, with the result that the service declines and the necessary extensions to distribution systems to serve the poor are not funded. An excellent description of the financial benefit for the poor in having a piped water supply, is given in a diagram by Arthur McIntosh [4] in which he shows how even with a necessary increase in tariffs, the poor are better off than they were when having to pay water vendors. I have slightly modified the diagram in Figure 3.2 to show that, even with the poor increasing their water consumption, which is inevitable when water is more easily available, they are still better off financially. My version of the diagram, which intends to show the principles, does not give costs or quantities, but for numbers in an Asian context, see McIntosh [4].

Low tariffs do not help the poor; on the contrary they deny them a decent water supply. So what is necessary to help the poor? Firstly, general subsidies should be phased out, with an associated increase in tariffs towards sustainable cost recovery, and available external funds should be directed to refurbishing and extending distributions systems. The refurbishment, coupled with effort to reduce all aspects of non-revenue water, will allow for continuous 24 hour a day supplies, and make water available for those yet to be connected. But refurbishment is expensive and takes time. It would be good if both the donors and the international banks would see such investment as a priority, because additional resources and additional treatment capacity cannot be utilised unless there is adequate distribution capacity.

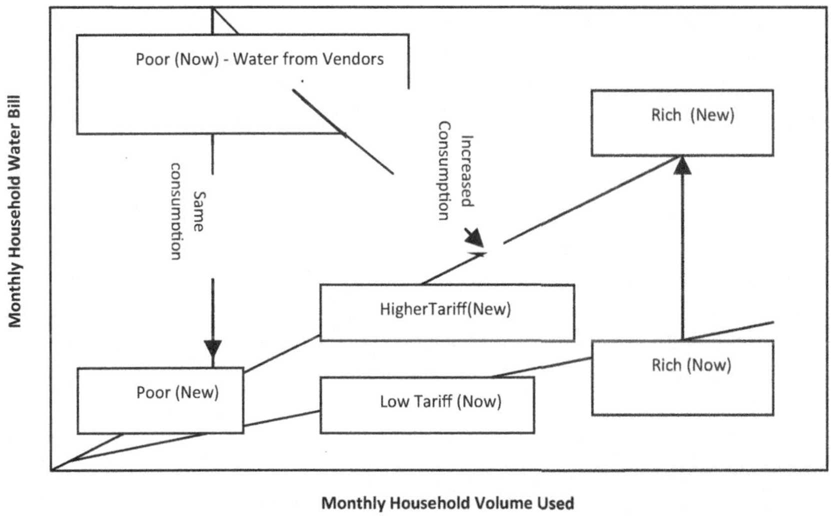

Figure 3.2 Increase tariffs to help the poor (*Source*: adapted from *Water and Poverty, Asian Water Supplies*, Arthur McIntosh, [4]).

Secondly, because the poor are unable to save money, there should be no connection charge, or it should be payable in easy instalments; see Chapter 8 for the sixty-instalment approach used in Chile. On installation, consumers should be given advice on how much tap water is used for various purposes, and the cost. But as shown in Figure 3.2, even with an increase in water usage, the bills the poor pay are likely to be lower than when they buy water at a high unit cost from water vendors.

Thirdly, the poor need assistance in how they pay their water bills. One approach is a "pay as you go" option. This is equivalent to the situation of paying for water from water vendors. One way of paying as you go is the use of pre-payment meters, an approach used in South Africa. It has been controversial, perhaps due to inadequate consultation in some cases and by post-apartheid reactions. It was seen [14] to have been introduced only in townships as a result of non-payment akin to the anti-apartheid rent boycotts of the 1980s, rather than being promoted as a means of helping the poor. Instead there was resentment that the "whites" did not have pre-payment meters. One good example of the successful use of pre-payment meters was in Mogale City [15] in South Africa. In 1998, the amalgamation of several local authorities resulted in a large increase in population requiring a water supply. The Local Council set up mass meetings with the communities, and distributed leaflets explaining why pre-payment meters were going to be fitted. Having gained community acceptance for the scheme, it was followed up with a door-to-door marketing campaign to encourage consumers to complete the application forms. In addition to pre-payment meters there is a lifeline tariff for water for the first 10 kls per month. The average consumption of water for those on pre-payment meters was less than 10 kls per month, so most of the water taken through pre-payment meters was bought at the lower block price. There was no charge to individual consumers for the meters or their installation. The system has been reported as highly successful from a water service provider point of view, with less than 1% of meters vandalised or bypassed, and payment levels up from 25% to 85%. In one community area, following the introduction of pre-payment meters, cost recovery from charges went from just 7% to 86%, based on operating costs. If the cost of capital for the meter installation is included, the improved figure was 45%. So not only were the consumers provided with a means of payment more suited to their circumstances, but through achieving greatly increased revenue, largely from the lifeline tariff rate, the community was some way towards achieving a sustainable water supply system. The question is whether the people of Mogale see it that way.

The water service provider in Cape Town is using a water management device as part of a water conservation and water management strategy. Its aim is promoted to the general public as saving water and helping to avoid people going into arrears on water payments. These devices allow water to flow from 4 am in the morning and turn off when the allocation for the day has been reached. The allocation for the day is generally based on 6 kls per month plus additional amounts to which householders commit for payment. A limited study [16] of residents with the device indicated inadequate public consultation and unreliability of the devices.

It is unfortunate and a barrier to progress that tools which could, with effective public consultation, provide benefits for the poor are so caught up in the controversial area of cut-off. The reality is that the poor do suffer unless there is sufficient revenue collection to achieve sustainable cost recovery. The examples above demonstrate the critical area of effective public consultation which is covered in Chapter 5.

An approach [17] which is being used extensively in Kenya and Uganda is mobile banking. Mobile phone operators are providing a utility bill payment service through mobile bank accounts. Users can buy credit for their accounts through local agents. As the transaction costs are low, utilities can provide for bills to be paid in frequent small amounts. The water consumers benefit as they avoid the inconvenience and cost of travelling to pay through banks or at water utility premises. The water utilities benefit through improved revenue collection and improved cash flow. This can lead to an improved water service and greater sustainability. There are prospects for these payment approaches to improve transparency and thus help address corruption. Greater security of revenue collection could improve prospects for external funding, and might even contribute to the development of advances in thinking on the design of future water supply systems necessary for rapid urbanisation. Important research on the use and benefits of mobile telephony in water management is taking place at the University of Oxford.

An alternative to a prepayment system is to subsidise the payment of monthly or quarterly bills. To do this effectively requires good local information to be able to operate a fair means-tested system. This is demanding administratively, with the usual problem of alienating those who are just outside the agreed definition of need. A successful system is in operation in Chile [18] and this is described in Chapter 8.

Another approach to assisting the poor is the use of block tariffs, with a low lifeline tariff for the first quantity of water used, which is part of the Mogale city approach described above. Block tariffs are better than a general subsidy for assisting the poor; however, the financial benefit is gained by all the consumers, including the more wealthy citizens. Achieving sustainable cost recovery with a lifeline tariff relies on those consumers who take more water being willing to pay more than its cost, to compensate for the under-recovery at lifeline tariff level. South Africa has a policy of introducing a lifeline amount of 6 kls per month at no cost. In Johannesburg a study (yet to be published) concluded that this achieved improved overall revenue collection, as the poorer consumers perceived it to be a fair system resulting in willingness to pay for the amount consumed above 6 kls a month. Israel [19] has a block tariff structure providing for 7 kls a month at the lowest level with roughly double the charge for each additional kl. For each family member over four people, an additional 3 kls is available at the lowest tariff. This requires good up-to-date information on household size.

The difficulty with block tariff systems is variability of household size; extended families, which are often in the low-income category, are penalised rather than assisted. Successful systems require something like the approach taken in Israel, with an additional lifeline amount based on household size, but there is a significant administrative burden in establishing and maintaining household records. Without good information on households, larger poor families are probably better off with a flat tariff system, providing that the bills can be paid in easy instalments.

THE CONTROVERSIAL ISSUE OF DEALING WITH NON-PAYMENT

Cutting off supplies, after due warnings, is commonly applied across the world. For example, in South Africa, government has taken the view that, without this threat, people will not pay their water bills, even with the benefit of a zero cost lifeline system. We know, as mentioned above, that the poor do find the money for water and pay a higher price than those connected to subsidised networks. The priority is how best to provide the poor with easy payment methods. The next requirement is how to implement a cut-off policy so that people do retain access to a safe water supply. In Durban South Africa [20] the first stage following non-payment is to fit a flow restrictor. If the outstanding bill is paid the restrictor is removed. If the bill is not paid the supply is cut-off but a street standpipe supply is made available to provide the free allocation of 6 kls per month.

In Kolhapur [21] in India, to achieve financial stability, the Mandal, the local community management, introduced a policy of disconnection to ensure that householders pay their bills. This was possible in this system, without accusations of penalizing the poor, due to the availability of public standposts. In the 18-year period up to 1999, only 16 disconnections had been necessary. In comparison, the other district- (ie non-community) managed systems in the same area of India have no such policy with a much higher level of non-payment and the risk of unsustainable systems.

In comparison with the South African and Kolhapur examples, the UK government has taken the view that no-one should be cut off, with the result that there are increasing numbers of non-payers. The implications of this are discussed in Chapter 9. In most cases if people use water wisely, water bills are a relatively small part of disposable income. Therefore we can assume that most of the non-payers are those who either won't pay or, without the threat of cut off, don't put aside enough money to pay for predictable water bills when they are received. I believe that some form of pre-payment system is the solution to this UK problem, to avoid an additional cost penalty on those people who do pay for their water services. However, an attempt to introduce pre-payment meters was challenged and deemed to be unlawful. I believe that legal provision needs to be made to address this deficiency.

One of the benefits of community-managed systems is the community pressure that is applied to achieve universal payment. Probably, the success in Kolhapur was partly as a result of community pressure, and not wholly associated with the threat of cut off. Similar benefits of community management were seen with the Savelugu system in Ghana (see Chapter 5 for brief descriptions of these examples).

RESPONSIBILITY OF DETERMINING TARIFFS

A number of the important policy issues have been discussed above and decisions on these are a matter for national or local governments. But having established policies, allowable cost recovery from charges should be set by an independent body which may be called a regulator, which is able to make a wholly objective assessment of operating costs and of the amount of infrastructure refurbishment required for long-term sustainability of the service. The independence of such a regulator can ensure transparency of the charge setting process, through effective consumer participation and through making key documents in the charge setting process available to the public. In this way, political interference in the implementation of policies designed for sustainability, is eliminated or reduced. It is not possible to have an objective method of establishing tariffs, to achieve sustainable cost recovery, without an integrated economic and water service objectives planning process.

THE NEED FOR AN INTEGRATED PLANNING PROCESS

A number of conditions are required to achieve sustainable water services. For water supply, there have to be reliable water resources based on defined extreme situations, the necessary water treatment processes to achieve the defined drinking water quality standards, and distribution systems that deliver water in the required quantities to the consumers without adversely affecting its quality. On sanitation, there should be an effective collection system and treatment system, whether or not there is a waterborne sewerage system or a community based recovery system, to achieve the specified environmental objectives. Increasingly, there are linkages between water supply and sanitation through integrated local systems. Whatever approach is taken, there should be clarity of policies, well-defined and realistic drinking water and environmental standards, and a planning system which allows all aspects to be considered in an integrated way.

In addition to new developments, there has to be consideration [22] of the maintenance and refurbishment of existing systems. So often when finance is limited, as indeed it generally is, there has been a tendency to delay investment on the old system, and to make do and mend. Ultimately, continuing deterioration of systems results in loss of serviceability, and failure to meet standards. The requirement is for operational measures that predict when there will be an impact on serviceability so that action can be taken in time. These actions become part of the operational plans.

All of the above aspects have to be financed and the costs recovered. This is why decisions on standards and system development should not be taken in isolation of financing requirements, or in isolation of the ability of the costs to be recovered through grants, subsidies and charges. There needs to be an integrated planning and funding process.

CITIES OF THE FUTURE

It is estimated by the United Nations [23] that the world's population will increase from 7 billion to 9.3 billion by 2050. Because of increasing urbanisation [24] most of the additional 2 billion plus will impact urban areas in developing countries. This has major implications for water resources and for integrated city and water and sanitation planning. Historically water services were part of cities' operations, but over recent decades water development has tended to be focussed on integrated river basin management. For example in 1974 water and sanitation services in England and Wales (see Chapter 9) were reorganised on a river basin structure. Concern over the equitable and effective use of river basins will increase, but there should be integration of water and sanitation provision planning with city infrastructure planning. Currently water networks often do not keep pace with growth in cities. An Oxford student in 2012 studied 29 rapidly growing cities in India and developed a conceptual model (yet to be published) that describes water network coverage, urban spatial characteristics, and urban growth patterns. Estimating population growth is difficult enough; determining the shape of that growth and what it means in water network design is more difficult. Much more research in this area is required. With water resource limitations and an increasing need for water re-use, a modular approach would appear to be appropriate. There has been progress on considering how water re-use can be integrated into water resource management. A "model" city approach is being adopted in Quingdao in China [25]. There is research work on "smart" network thinking [26] to enhance existing networks. I would like to see much more research on water network "spines" within which the more local network thinking could be incorporated. In 2006, the International Water Association (IWA), recognising the impact of rapid urbanisation, made Cities of the Future [27] one of its key themes. IWA has a major ongoing programme of activity in this area. See important publications on cities of the future ([28], [29]).

Singapore [30] is an excellent example of the integration of water and wastewater planning as part of city development, and of adopting a sustainable development approach adopted long before that term became widely used. Singapore, although not having to cope with the potential for city sprawl due to its size and natural boundaries, has had to develop innovative ways of capturing fresh water and of re-use. Although less recognised than its water resource developments, the planning approach to the management, maintenance and extensions to its distribution systems has been a key part of achieving sustainable continuous (24/7) supplies to all of its citizens.

SOME EXAMPLES SHOWING THE NEED FOR INTEGRATED PLANNING

The World Bank North China Study [31] included consideration of the institutional governance and regulation aspects of water services in China. A sample of five cities was taken, to give a range of institutional situations and levels of progress in implementing national governments' required reforms as part of the so-called marketization process. Two of these cities, Jinan in Shandong Province and Taiyuan in neighbouring Shanxi Province, had severe water finance problems arising from a lack of integrated planning

In Jinan, in order to protect the city's title of "City of Springs", the Provincial Government took the decision to stop groundwater abstraction and to build two reservoirs fed by the Yellow River. This was a laudable aim but it was done without effective discussions with the city of Jinan which had to bear the cost of the reservoir developments. The reservoir capacity far exceeded the water resource requirements of the city, thus incurring much higher capital cost than was necessary, leaving Jinan Water Supply Company with a high level of debt.

There was a similar situation in Taiyuan with the Shanxi Provincial Government's decision to embark on a long distance transfer project from the Yellow River. This was necessary due to the need to resolve the problems of over-abstraction of groundwater, but the transfer level greatly exceeded the supply needs of the city. Also, Taiyuan Waterworks Company has been unable to implement the closure of the large abstraction wells providing the major groundwater consumers, as they are regulated at national or provincial level. Both planning and implementation, together with funding and cost implications needed to have been handled in an integrated way. The World Bank report recommended, and I quote, "governments at all levels should ensure that the reform of the structure of the water and wastewater sector is carefully designed and implementation is well planned and, in particular:

- Develop coherent frameworks for government functions of policy, planning, regulation, asset ownership and business ownership. This must include the separation of policy, ownership, operations and regulation.
- Clarify the responsibilities of different agencies and develop agreements and inter-agency procedures for cooperation, particularly for integrated planning across the urban water cycle. Establish coherent and clear lines of responsibility and accountability."

It is common in many countries to rely on effective liaison between various government levels rather than to establish an integrated planning and regulation process. For example, in the United States, standards are set by the Federal Government, environmental standards are regulated at state level, but responsibility for meeting the standards rests with the cities. As a consequence many cities do not see it as their responsibility to be self-funding, and seek federal funds for implementing new regulations and for replacing ageing infrastructure. This attitude has been created partly by federal funding of the introduction of secondary wastewater treatment in earlier years, creating a subsidy mentality. Changing funding policy would require a substantial increase in water service charges, which would be unpopular, even though generally charges are currently very low and the required increases would not result in high household costs in relation to average income. See examples of successful transformations to sustainable cost recovery in the United States in Seattle (Chapter 7) and Atlanta (Chapter 8).

Governments tend to set drinking water quality standards without regard to how they are going to be achieved. In many countries in Africa, the current quality of drinking water is well below the standards which have been set. The standards are usually based on the WHO guideline values, which is sensible, and countries should aim to achieve those values as a critical public health measure. However, there should be recognition that compliance will take time and require careful planning and investment. There is no point in having standards unless there is effective monitoring, and no point in having monitoring unless there is an enforcement process and the means to plan and fund the necessary improvements. In such cases, there should be recognition of the need for interim standards within a long term planning process, so that those involved can see a clear way ahead, and can measure and celebrate progress as standards are met.

ENGLAND AND WALES

One of the best examples in the world of an integrated planning process is the one adopted as part of the regulatory system in England and Wales, and in place at the time of writing in 2013. This has created clarity on delivery requirements, a method of planning for meeting both quality standards and asset refurbishment, and an effective means of establishing the necessary funding of improvements and of ensuring delivery of commitments. It is not integrated with city planning but this is not critical as the major urbanisation took place during the industrial revolution, around the middle of the 19th century. Chapter 9 is the story of England and Wales, covering all aspects, including the important history, but it is worth summarising the integrated planning process in the context of this chapter. The integrated planning and price fixing process is called the Periodic Review, with the period of review being at five-year intervals. It brings together all aspects of water company delivery over five-year periods and the average limits on charges that the water companies can make each year. This is a transparent process with all the key government and regulator documents being in the public domain. An outline of a typical sequence of events is as follows:

(1) Ofwat, the economic regulator, publishes the timetable of events.
(2) Consumers are consulted on what improvements they would like to see and how much extra, if any, they are prepared to pay for them.
(3) Government publishes, in outline, the quality improvements it would like to see achieved during that period.
(4) The Drinking Water Inspectorate (DWI) and the Environment Agency (EA) prepare guidance for the water companies on what is required to achieve the desired quality improvements.
(5) The water companies submit their proposals to DWI and the EA on how they will meet the requirements, and the two quality regulators say whether they are appropriate.

(6) The water companies submit business plans giving the improvement schemes and their costs.
(7) Ofwat considers these plans and whether the costs are realistic and, taking into account both the costs and the required efficiency improvements, calculates the allowable price limits to finance the programmes of work.
(8) These price limits are then reviewed, including by Government, who consider whether the quality improvements are affordable or whether some aspects of quality improvement should wait until a later period.
(9) There are then discussions between Ofwat and each company before Ofwat announces its decision on the price limits for each company.

The plans become water company commitments for each five-year period which are monitored and if necessary enforced by the regulators. This process has delivered clarity of objectives and has given the water companies certainty on what is required of them, and delivered a greatly improved and more efficient service for the consumers.

REFERENCES

1. Lights! Water! Motion! Booz, Allen and Hamilton, 2007.
2. Sustainable cost recovery. International Water Association Reference Paper. IWA, website www.iwahq.org.uk
3. Water Supply (2004). Drainage, Sewerage and Sanitation Management Programme in Haiphong 2001–2004 Project Completion Report, Finnida, December 2004.
4. McIntosh A. C. (2003). Asian Water Supplies: Reaching the Urban Poor. Asian Development Bank, 2003. ISBN: 9781843390435.
5. Farley M. and Trow S. (2001). Losses in Water Distribution Networks: A Practitioner's Guide to Assessment, Monitoring and Control. IWA Publishing. ISBN: 978190022112.
6. World Population Prospects: The 2010 Revision. Population Division of the Department of Economic and Social Affairs of the United Nations Secretariat.
7. World Urbanisation Prospects: The 2011 Revision. Population Division of the Department of Economic and Social Affairs of the United Nations Secretariat.
8. PAAB Annual Report, www.paab.my
9. Sewerage Rehabilitation Manual (2001). 4th edn, WRC, Swindon, April 2001. ISBN: 9781898920397 (also online at http://srm.wrcplc.co.uk).
10. Asset Management: A handbook for Small Water Systems (2003). One of the Simple Steps for Effective Performance (STEP) Guide Series, Washington DC, US EPA 816-R-013-06, September 2003.
11. Asset Management Planning and Reporting Systems for Water Utilities (2006). AwwaRF Report 91095F. IWA Publishing, November 2006. ISBN: 9781843399513.
12. Savic D. A., Djordjevic S., Dorini G., Shepherd W., Cashman A. and Saul A. (2005). COST-S: a new methodology and tools for sewer asset management based on whole life costs. *Water Asset Management International*, **1.4**, 20–24.
13. Sharma S. (2002). Water Markets Exclude the Poor. World Summit on Sustainable Development, Johannesburg.

14. von Schnitzler A. (2008). Citizenship prepaid: water, calculability, and techno-politics in South Africa. *Journal of Southern African Studies*, **34**(4), 899–917.
15. Rabe M. (2001). Experiences in prepayment meters. Based on a paper given at vending options 2001. *ESI Africa.*, **2**, 2001.
16. Pereira T. (2009). Household Impacts of Water Management Devices – Summary of Initial Research. Presented at EMG Water and climate change seminar, Cape Town, South Africa, 26 November 2009.
17. Foster T., Hope R., Thomas M., Cohen I., Krolikowski A. and Nyaga C. (2012). Impacts and implications of mobile water payments in East Africa. *Water International*, doi: 10.1080/02508060.2012.738409.
18. Bitran G. A. and Valenzuela E. P. (2003). Water Services in Chile: Comparing Private and Public Performance. World Bank, 2003. Paper available on website http://rru.worldbank.org/Viewpoint/index.asp
19. Fernandes G. (2012). Economic Aspects in Water Management in Israel – Policy and Prices, www.water.gov.il/Hebrew/ProfessionalInfoAndData/2012/06-Israel-Water-Sector-Economics-(Policy-and-Prices).pdf
20. Water Policy Durban Imanage 51 517: Water policy summary 24 May 2002.
21. Sustainable Community Management of a Multi-Village Water Supply Scheme in Kolhapur, Maharashtra, India. Water and Sanitation Program Field Note SPI Series 2 DFID, 2000.
22. Rouse M. (2013). Policy brief: the urban water challenge. *International Journal of Water Resources Development*, doi: 10.1080/07900627.2013.791568.
23. UN (United Nations) (2011). World Population Prospects: The 2010 Revision. United Nations, New York.
24. UN (United Nations) (2012). World Urbanisation Prospects: The 2011 Revision. United Nations, New York.
25. Novotny V. (2011). Holistic Approach for Distributed Water and Energy Management in the Cities of the Future. IWA International Conference. Cities of the Future Xi'an Technologies for Integrated Urban Water Management. September 15–19, Xi'an, China.
26. Vairavamoorthy K. (2012). Networks of the Future (Smart, Multipurpose and Flexible by Design). Paper presented at the World Water Congress and Exhibition, September, 2012, Busan, Korea.
27. Profile of IWA Theme: Cities of The Future. http://www.iwawaterwiki.org/xwiki/bin/view/Organizations/+Cities+of+The+Future
28. Novotny N. and Brown P. (eds) (2007). Cities of the Future: Towards Integrated Sustainable Water and Landscape Management. ISBN: 9781843391364.
29. Hao X., Novotny N. and Nelson V. (eds) (2010). Water Infrastructure for Sustainable Communities. IWA Publishing, London, ISBN: 9781843393283.
30. Tortajada C., Joshi Y. and Biswas A. K. (2013). The Singapore Water Story: Sustainable Development in an Urban City State. Routledge, London.
31. North China Water Quality Management Study Program. (2005). Component E: Regulatory and Institutional Study. World Bank (WB) and the Ministry of Construction (MOC), China.

Chapter 4

Drinking water quality and regulation

Better safe than sorry

In the developed world today, waterborne-induced illness is rare and there are few deaths from waterborne disease, but this was not the case before water treatment and safe handling of human excrement was introduced at the end of the 19th century. The cholera outbreak in London of 1854, with the pioneering epidemiological work of Dr John Snow, was an important step in the understanding of waterborne disease. A book [1] by Steven Johnson adds to the information on the arguments about the nature of the transmission of infection at the time of the outbreak. However, in the developing world today, even though it is understood how and why, poor access to safe water and inadequate sanitation is a major cause of illness and deaths. According to Global Health Observatory, WHO, in its report of 2004, there were 1.9 million deaths per year from diarrhoea of which 88% were attributable to unsafe water, inadequate sanitation and lack of hygiene [2]. Higher and lower numbers are quoted. Whatever the true figure, waterborne disease results in a large number of deaths each year. Those most affected are children under the age of five in developing countries.

Standards need to be designed to meet citizens' requirements relating to drinking water safety and aesthetic quality, taking into account logistics and affordability. In the developed world there is a tendency now to make standards unnecessarily stringent, and in the developing world to adopt standards that are unrealistic in the short to medium term. Almost everywhere too little attention is given to monitoring and enforcement.

I stress at the beginning that there is no point in having standards unless they are monitored, there is no point in monitoring compliance with standards unless it is followed by enforcement, and no point in having an enforcement process unless there is a means by which the necessary improvements can be funded and implemented. This chapter tracks the steps from threats to health leading to standards, to current thinking on how best to achieve safe drinking water, to monitoring and enforcement, to sanctions for failures. The chapter finishes with the example of the Drinking Water Inspectorate in London as one good model of a drinking water quality regulatory body.

THREATS TO THE SAFETY OF DRINKING WATER

The threats can be considered under three headings, namely microbiological organisms, chemical substances and radionuclides, with pathogenic microbiological organisms being by far the most important. It is difficult to give accurate figures on numbers of cases of illness and deaths from pathogens but estimates give cases in billions and deaths in millions each year. Figures taken from various WHO documents are given in Table 4.1. There is a tendency by lay people to see the main threat as cholera, presumably because such outbreaks hit the headlines. The figures for cholera vary considerably year by year (Figure 4.1) but its impact although important is only part of the diarrhoeal disease total from many different organisms. Table 4.1 includes the number of cases of the illness giardiasis. It is estimated [3] that around one third of the population in the developing world has been affected by giardia. It is the most common cause of waterborne illness in the United States [4], but there are few deaths. In the developed world protection against *giardia* and *cryptosporidium* has become a major consideration due to these organisms being resistant to chlorine.

Table 4.1 Incidence and deaths from waterborne disease.

Water, sanitation and hygiene related pathogens		
Disease	**Cases per year (thousands)**	**Deaths per year (thousands)**
Total Diarrhoeal Disease	2,000,000	2.400
Cholera	3000–5000	100–120
Giardiasis	2000	low

Sources: GBD2004 and CDC.

The Global Burden of Disease Study, GBD10, [5] published in December 2012, described as the largest systematic effort on the subject, compared the situation in year 2010 with that in 1990. The study concluded "The results show that infectious diseases, maternal and child illness, and malnutrition now cause fewer deaths and less illness than they did twenty years ago". Table 4.2 gives the figures from that study on deaths from total diarrheoal diseases, cholera and cryptosporidiosis. At the time of writing the data and the methodology used in GBD10 are being debated.

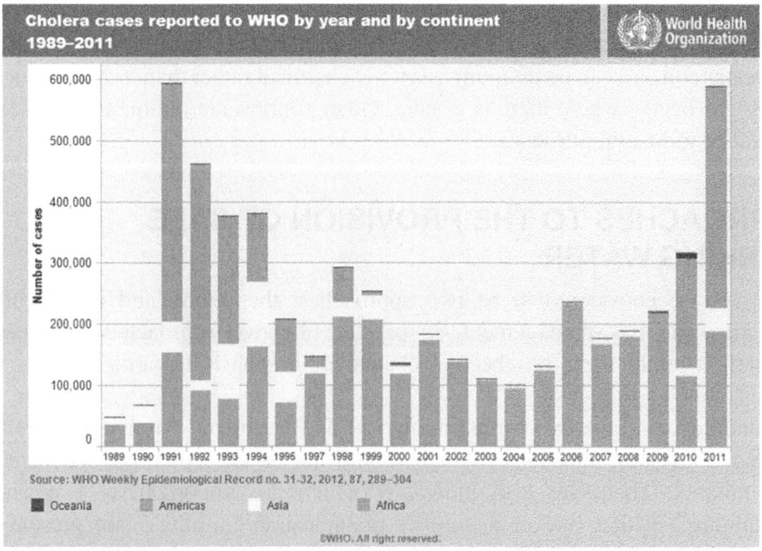

Figure 4.1 Number of cholera cases reported from 1989–2011 (*Source*: WHO).

Table 4.2 Comparison of deaths in 1990 and 2010.

Estimated worldwide deaths 1990 and 2010 (thousands)		
Disease	**1990**	**2010**
Total Diarrhoeal Disease	2487	1445
Cholera	120	58
Cryptosporidiosis	222	99

Source: GBD10.

In comparison with microbiological organisms, there is less risk from chemical substances which occur rarely at toxic levels. Chemicals are considered in two groups, inorganic and organic. Generally, the concern with inorganic chemicals is long-term effects with standards being set based on a lifetime of consumption. There are cases where there are short-term impacts of inorganic chemicals, such as from high levels of arsenic in some groundwaters, and it is important that bottle feeds for infants are not made from water containing high levels of nitrate, because of the risk of methaemoglobinaemia (blue baby syndrome). The situation with organic chemicals of which there are many thousands of substances (some of which occur naturally and others are manufactured), is a little different, as can be seen below in the approach to the setting of standards.

The third group, that of radionuclides, is rarely a problem and insignificant compared with the risks from pathogens. Most of the radionuclides occur naturally in the environment. Examples are uranium and radon. Uranium levels can be a concern when water from flooded mines contaminates water courses, although the greater concern can be uranium as a toxic chemical rather than as a radionuclide. Radon gas occurs where there is granite. Other sources are radionuclides used for medical or industrial purposes.

APPROACHES TO THE PROVISION OF SAFE DRINKING WATER

There can be considered to be two approaches, the setting and monitoring of standards, and risk management. In practice the two approaches go together, although recently there has been a change of emphasis towards effective risk management.

Standards can take two forms, numerical and treatment. Numerical standards, which are levels of substances which should not be exceeded, are set for actual contaminants. There are also indicators which are important as a means of determining whether one or a number of undesirable "things" are present. An example is *e-coli* which, although not a pathogen itself, it is an indicator of faecal contamination in which certain pathogenic organisms could be present. The presence of *e-coli* indicates the need for a thorough investigation.

The requirement to disinfect water intended for human consumption is one example of a treatment standard. Treatment standards can be based on the effectiveness of specific treatment processes, and can be expressed in numbers of log (logarithmic) removal. A one log removal process will reduce the quantity of "a particular thing" to one tenth, two log removal to one hundredth, three log removal to one thousandth, four log removal to one ten thousandth, and so on. Setting standards on this basis requires an assumption to be made that the numbers of something remaining will not constitute an infective dose. There can be alternatives of a treatment standard or a numerical standard, as in the UK approach which was used to protect against the organism *cryptosporidium*. This is discussed later in this chapter.

The recent change to give greater emphasis to risk management arose from concern over the tendency for more and more numerical standards. For example, in the United States, there was a legal requirement to include additional substances periodically. A number of people closely involved in drinking water quality said that there must be a better way of approaching drinking water quality, because although numerical standards are important, measurement of quality at the point of delivery is too late, other than for verification. As a result of this concern, the first of two very important international workshops was held in Bonn, Germany in October 2001. At this workshop there was a fundamental review of the approach to the delivery of safe drinking water. This led to a change of emphasis in the revision and development of The WHO Guidelines for Drinking-Water Quality. There was a second Bonn workshop in February 2004 which led to the Bonn Charter. The Bonn Charter [6] and the WHO Guidelines [7] are interrelated and complementary documents. They were launched together in September 2004. They both refer to the need for "water safety plans" as a disciplined approach to risk management. Although I prefer the term "drinking water safety plan", as it makes it clear that the reference is to drinking water, in this book I use the term "water safety plan" for consistency with WHO terminology. I will return to setting numerical standards later in the chapter.

THE BONN CHARTER

The Bonn Charter is a set of high-level principles which can be applied worldwide for the supply of safe drinking water. It is important to recognise that the Charter is concerned with drinking water quality, although many of the principles apply equally to all aspects of water supply. Figure 4.2 gives a diagrammatic representation of the Bonn Charter framework. The important goal is described as "Good safe drinking water which has the trust of consumers". There are some key words here. "Safe" means safe, especially safe from pathogens. The interpretation of "good" has to take into account in which country the definition is being applied as the aesthetic quality depends critically on affordability. The word affordability does not appear in the Charter, although there is reference to the need for standards to reflect the progressive realisation of goals. I make reference to affordability, as we have to consider the world as it is, not as ideally we would wish it to be. In this context, I discuss below the need for interim achievable standards. Returning to the goal, the word "trust" is important, reflecting the need for good public participation, and the importance of full reporting of drinking water quality.

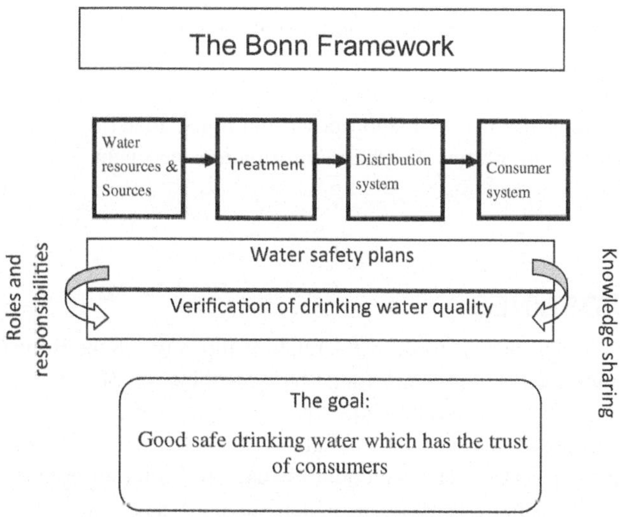

Figure 4.2 The Bonn Framework (*Source*: The Bonn Charter).

The Charter aims to provide the various stakeholders, politicians, government officials, regulators and senior water utility managers, with a framework for the development of an integrated approach to the delivery of safe drinking water, giving some key principles and listing roles and responsibilities. Some of the key principles are:

- there is a need for an integrated approach to include all organisations involved from catchment to tap;
- there should be close co-operation between all in the supply chain;
- transparency is vital for consumer confidence;
- drinking water quality standards, at consumers' taps, will depend upon local circumstances, but should cover both safety and acceptability;
- the need for clarity of responsibilities for each part of the supply chain.

The Charter considers the roles and responsibilities of the various stakeholders. Among the key responsibilities for governments are the creation of independent systems of coordinated water quality regulation, and the provision for institutions and coordination arrangements to deal with the risks to water quality from catchment to tap. Key responsibilities of water service providers, the utilities, include taking the lead, and working with others in the development and implementation of water safety plans. The development of water safety plans is covered in a section below. Another responsibility of utilities is ensuring that the full cost of service provision is identified, including the cost of maintenance and provision of assets. This is another context in which the need for full-cost recovery is important. In addition to the core responsibility of regulators in relation to drinking water quality standards, stress is given to the need for transparency and the full reporting of compliance results in a way which can be understood by all. Consumers themselves are regarded as having responsibilities in preventing contamination of drinking water in houses, including avoiding the use of inappropriate plumbing materials. Importantly, the Charter implores all parties to share information pertinent to achieving safe drinking water.

WHO GUIDELINES

At this stage it is appropriate to show the relationship between the Bonn Charter and the WHO Guidelines, which is illustrated by a combination of Figures 4.3 and 4.4. Figure 4.3, taken from the WHO Guidelines, gives the structure of the guidelines, showing the emphasis given to "water safety plans" within the central framework for safe drinking water. The guideline values for microbiological, chemical, radiological and aesthetic aspects, on which standards are based, are now shown as supporting information.

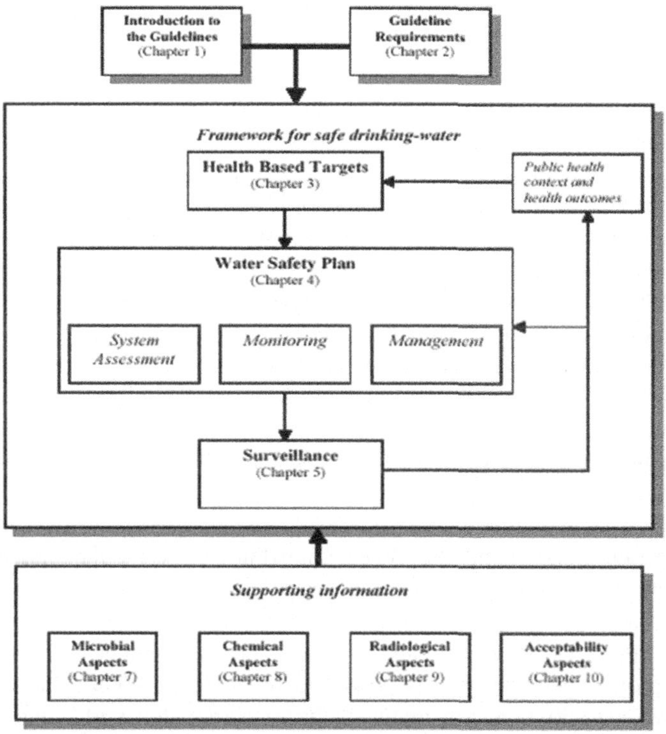

Figure 4.3 WHO Guidelines Structure (*Source*: Guidelines for Drinking Water Quality, WHO 2004).

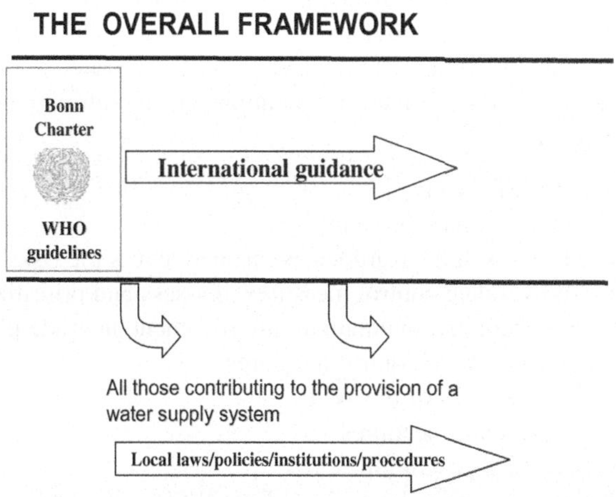

Figure 4.4 Integrated International Guidance (*Source*: Presentation at Launch of Bonn Charter 2004).

The approach includes the setting of health-based targets. Although this is highly desirable, with the aim of setting investment in water services against health needs, and measuring the health improvement impact of that investment, I have concerns about the practicality of the approach, especially in developing countries. It is known from a WHO study [8] that the benefits from investment in water and sanitation in developing countries greatly exceed the costs. Also, obtaining adequate data on which to base the health targets is difficult and time consuming. The process of developing drinking water safety plans, which in many cases can be done very simply, identifies both investment and system management priorities and, in my view, health-based targets are not necessary to decide that those identified actions are worthwhile. There is also the risk that the effort taken in producing health-based targets could delay the preparation and implementation of water safety plans or, even worse, deter governments from embarking on their development.

The Bonn Charter and the WHO Guidelines provide international guidance, but their implementation involves the incorporation of the recommended approaches, together with the guideline values, into local laws, policies and procedures. With both the Bonn Charter and the WHO Guidelines stressing the importance of water safety plans (WSPs), it is appropriate at this stage to begin with guidance on the preparation of plans. Chapter 4 of the WHO Guidelines gives detailed advice on the development of plans with reference to a number of examples. A book [9] providing guidance was published by WHO in 2005. Since that time WSPs have been prepared in many countries across the world and extensive guidance has become available, much of which is incorporated in the Water Safety Portal [10]. In 2012 WHO published guidance on water safety planning for small community water supplies [11]. The Water Safety Portal provides guidance on preparing WSPs, provides tools to assist in their development and gives some case studies, and provides support to Water Safety Plan Networks covering Africa, and the Asian-Pacific and Latin American-Caribbean regions. The Portal gives 11 modules in the preparation and monitoring/audit of water safety plans. They are:

- Assemble the WSP Team
- Describe the water supply system
- Identify the hazards and hazardous events and assess the risks
- Determine and validate control measures, reassess and prioritize risks
- Develop, implement and maintain an improvement/upgrade plan
- Define monitoring of the control measures
- Verify the effectiveness of the WSP
- Prepare management procedures
- Develop supporting programmes
- Plan and carry out a periodic review of the WSP
- Revise the WSP following an incident

WATER SAFETY PLANS

Readers wishing to embark on the preparation and implementation of WSPs should study the advice and experience available on the Portal and elsewhere. Based on my own experience I discuss some of the steps, consider audit of WSPs, and refer to WSPs in developing countries including their use as a means of informing improvement programmes.

The boundaries of a system can depend upon the source, the area supplied by a treatment works, or the system supplying a defined distribution system. The choice can be for convenience, or because a particular supply system is susceptible to some risks of known concern, or to reflect a team's responsibilities. A source can be a reservoir or river catchment, an aquifer, or a desalination plant, or a reuse source or, as in many cases, combinations of individual sources. Based on the choice and definition of the system being covered, a project team is formed. The team should include members from each organisation in the supply chain covering the water source, water treatment, water distribution and, if possible, consumers' premises. Generally, the team should include key people from a number of different organisations; this requires the management of those organisations to recognise the importance of collaboration and sharing of information. Drinking water safety is far too important for turf wars to get in the way of teamwork, not only in the development and operation of water safety plans, but also in the operation of systems in the future. The team leader is most likely to be a member from the water service provider, but the choice can depend upon circumstances, and on the availability of key individuals.

A good starting point for the team is an initiation workshop facilitated by someone with experience in the process of producing water safety plans. A workshop provides the opportunity for the team members to get to know each other, and to understand each other's challenges. It is desirable to invite others, who can provide valuable inputs on the supply history and potential risks to supply, and for there to be management involvement to ensure a commitment to implementation. At this stage it is important to assemble available information and documents on the system under study, such as an accurate map and flow diagram of the treatment works, and to identify deficiencies in any of the information. One output from the workshop should be an agreed timetable for the project, with a firm date for the completion of the first (risk assessment) stage.

The risk assessment involves each member of the team identifying the hazards to the safety of drinking water in his or her part of the supply chain. Hazards can be from a number of sources: environmental including pollution incidents, system failure, or from deliberate attempts at contamination by man. Dealing with risks from sabotage, terrorism or vandals is often seen as a separate requirement, but I consider it important for these to be included in one drinking water safety plan, as all aspects need to be built into operational procedure and reaction planning. At this stage I have used the words "hazard" and "risk" rather loosely. The Oxford English Dictionary does not help very much but there is a useful discussion [12] of the distinction between the two words by the UK Centre for Occupational and Environmental Health. COEH refers to hazard as the potential to cause harm, and risk as the likelihood of harm (in defined circumstances, and usually qualified by some statement of the severity of the harm). This highlights two important measures, likelihood and severity, which need to be considered together, and the WHO Guidelines provide a simple scoring system (Table 4.3) to assist in identifying the relative importance of risks. Such a simple system should be used only as a guide because, for example, the catastrophic nature of a hazard may be such that, however unlikely it is to happen, it needs to be taken very seriously. The risk-scoring table uses the following definitions which are useful, but again they should be regarded only as giving some guidance and not used prescriptively:

Likelihood

- Almost certain – once a day
- Likely – once a week
- Moderate – once a month
- Unlikely – once a year
- Rare – once every five years

Consequences

- Catastrophic – potentially lethal to a large population
- Major – potentially lethal to a small population
- Moderate – potentially harmful to a large population
- Minor – potentially harmful to a small population
- Insignificant – no impact or not detectable

Table 4.3 WHO suggested risk scoring; a simple risk scoring system.

Likelihood	Severity of consequences				
	Insignificant	**Minor**	**Moderate**	**Major**	**Catastrophic**
Almost certain	5	10	15	20	25
Likely	4	8	12	16	20
Moderate	3	6	9	12	16
Unlikely	2	4	6	8	10
Rare	1	2	3	4	5

Source: Water Safety Plans The World Health Organization 2005.

Each hazard and its risk assessment should be recorded, together with any supporting information. Table 4.4 is an example of a recording system.

Table 4.4 Table for recording risk assessment information.

Part of supply chain	Source/cause	Hazard	Preventative and control measures	Likelihood	Severity	Risk	Downstream controls	Comments and further considerations

Once the risk assessment has been completed, each member of the team should review the controls in place to counter the hazard. Is there evidence that the controls are working? Have there been incidents where the controls have been inadequate? What preventative measures could be taken to reduce the risk? Could improved or additional downstream (i.e. later on in the supply chain) controls be put in place? To give an example, in the case of a hazard of a river pollution incident, there should be preventative measures, such as bunds on a factory premises to contain a chemical spill. A downstream measure could be detection of pollution in the river, coupled with the closing of the intake to the treatment works and, if necessary, additional treatment at the works. In another case, a downstream control for a failure of disinfection at a treatment works could be chlorine injection into the distribution system. The analysis of control measures is likely to identify some priorities for improvement which should be acted on at this early stage, thus giving some early benefit for the work done on the project. Although the control improvements will reduce the likelihood of an incident taking place, they will not change the consequences should it do so.

The next step is to set both target levels and action limits for each of the controls, and to record the designated reaction measures should an action limit be breached. The target levels on controls at a treatment works are likely to include the drinking water quality standards at which, in general, noncompliance can be dealt with under routine conditions. An action level should be set at a figure at which immediate action needs to be taken. At the beginning of the supply chain, at the source end, some of the target levels are likely to be more subjective and could require longer timescales for achievement. For example, it may require a 10-year or more programme to "clean up" a catchment. In such cases, the action limits need to be set to give as early a warning as possible for downstream reaction programmes to be enacted. Monitoring against the limits is critical, and it is necessary to define what, when (frequency) and by whom. The other necessary component is a means of verifying that the corrective action has worked. The outcome of this work can be recorded as in Table 4.5 below.

Table 4.5 Record of action limits, monitoring and corrective actions.

Hazard	Critical limits		Monitoring			Reaction steps	Verification
	Target level	Action Level	What	When	Who		

The above provisions should be converted into operational procedures, then operators should be trained in the new or revised procedures; the management responsible for each part of the supply chain should establish a means of checking that the procedures and training do result in an effective system for the supply of safe drinking water supply. Exercises which try to simulate real incidents provide a useful means of checking the effectiveness of the systems, especially the critical aspect of communication between the various organisations involved from catchment to tap.

BEYOND THE IMMEDIATE CONTROL OF THE WATER SERVICE PROVIDER

The Bonn Charter makes it clear that hazards have to be considered from source to consumer. Some of the greatest risks may exist once the water has left the water service provider system whether that be a piped system or a well. In many parts of the world final distribution relies on the informal sector delivering water by road tanker or by cart. Water may have to be stored in tanks or multiple containers. Such water can be vulnerable to contamination. WHO provides advice on protecting drinking water in households [13]. Devices for treating water in premises are available but these present hazards if they are not maintained. WHO and UNICEF have published a toolkit [14] for monitoring and evaluating household treatment and safe storage programmes. The toolkit refers to water safety plans as a useful instrument in determining when household systems are appropriate. The importance of hygiene measures such as hand washing and covering water storage containers is stressed. Consumer and Community Groups have a role in getting people to understand the risks to health and how they can be managed. Advice to householders should be provided in an easily understood form. This may include leaflets giving advice in illustrations for the illiterate. With concern over the risks of cryptosporidiosis to the immune-compromised, WHO issued a bulletin [15], advising on household control measures.

GOVERNANCE AND REGULATIONS

The extent to which the emphasis in the delivery of safe drinking water has moved towards managing risk is reflected in intentions for future regulations, and in how regulators [16] are adopting the thinking. During December 2006/January 2007 the International Water Association (IWA) held an internet forum on the Bonn Charter; those taking part described how the Charter and drinking water safety plans were being taken forward in a variety of ways, depending on countries' approaches to water management, standards and regulation. There were indications by the European Commission that water safety plans should be encouraged, or even incorporated, in the next EU Drinking Water Directive. Increasingly water safety planning is being adopted in policy and embedded in regulations. I give a few examples.

The New Zealand Health (Drinking Water) Act 2007 requires the establishment of public health management plans for all supplies whether urban or rural. Supply units serving less than 5000 people can obtain training and assistance in writing plans through the Technical Assistance Programme. Assessors appointed by the Director-General of Health check whether suppliers are in compliance with the Act. There is enforcement via an escalating series of penalties for non-compliance.

The Singapore drinking water regulations, the Environmental Public Health (Quality of Piped Drinking Water) Regulations, require water suppliers to develop and operate water safety plans to a specified standard. The requirements are set sown in the Code of Practice on Piped Drinking Water Sampling and Safety Plans. The Drinking Water Inspectorate audits the plans and their operation.

The regulator in Ghana, the Public Utilities Regulatory Commission (PURC), responsible for drinking water quality piloted an approach to water safety planning and provided initial training for the roll out of WSPs across urban supplies. Procedures for the audit of the plans are built into the PURC Inspectors' Manual. The pilot project is discussed later in this chapter.

In South Africa [17], as part of drinking water quality incentive-based regulation, water safety and incident response management plans are required. The certification system, named blue-drop, specifies risk assessments of catchment, treatment works and reticulation systems. A paper [18] published in 2006 had provided the framework for a risk management approach.

In India policy issues and governance aspects [19] were considered in the development and application of water safety plans for rural areas. The work addressed a number of key elements for successful application including essential technical support to the Gran Panchayats (GPs, the local parliaments) and their Village Water and Sanitation Committees (VWSCs) who are responsible for the water services. WSPs were seen as an important vehicle in developing improvement programmes, with the need to integrate them into the planning processes. Each layer of government has a role to play to support local operations, as demonstrated by Figure 4.5, the inverse pyramid taken from the Water and Sanitation Programme (WSP) document.

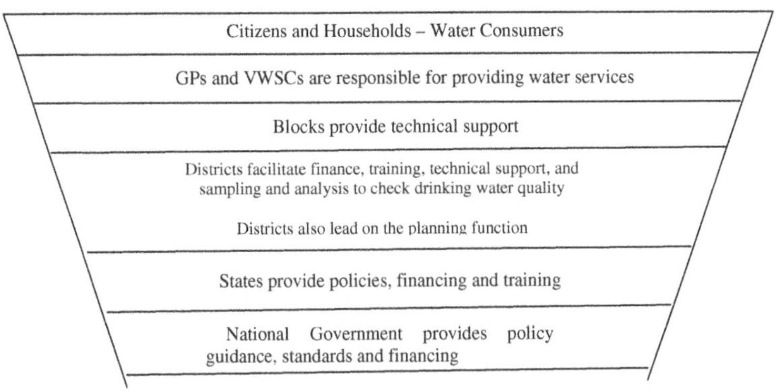

Figure 4.5 Inverse Pyramid of Support (*Source*: WSP Delhi).

The drinking water regulations in England and Wales make specific provisions for drinking water safety and require water companies to implement a risk (water safety plan) approach to water production and distribution. The regulations are enforced by the Drinking Water Inspectorate [20]. The Scottish Water (Objectives for 1st April 2010 to 31st March 2015) Directions 2009 [21] issued by Scottish Ministers required the water company supplying the whole of Scotland to prepare water safety plans for all supplies during the objective period.

In Hungary [22], by Government Decree 201/2001 which came into force in 2009, all water supplies serving more than 5000 people are required to have water safety plans with progressive implementation by July 2014. Some emphasis is given to risk management of drinking water resources.

An excellent example of regional co-operation is in the Pacific Island Countries who came together to develop an approach to drinking water safety through the introduction of sound policies and water safety plans. The resulting report [23] is not only a valuable account of the approach but it also provides a good manual on water safety planning.

AUDIT OF WATER SAFETY PLANS

Audit approaches can be split into three groups, self-audit, third party accreditation, and audit by a regulator. The most common approach in the developed world is a HACCP-type accreditation system. HACCP (Hazard Assessment and Critical Control Points) is used extensively in the food industry and this formal approach to risk assessment has been adopted in Australia. The approach in Australia ([24] and [25]) has been driven by the requirements of the National Health and Medical Research Council Framework for the Management of Drinking Water Quality (DWQMF). These requirements have been incorporated in the Australian Drinking Water Guidelines [26], which are based on a combination of HACCP and ISO 9000. The Framework gives a quality management system to be followed by water utilities using a catchment to tap risk assessment approach. Using approaches developed for the food industry, there is third party accreditation of the operational procedures and, if deemed by the assessor to meet the required accreditation standard, the water utility is granted a HACCP Certificate.

For some systems, for example in Melbourne Australia, the scrutiny of water safety plans has been taken one step further through using a Framework Continuous Improvement Tool (FCIT) [27], which has been developed by the Water Services Association of Australia. FCIT is a tool to judge the robustness of a water safety plan. It involves asking four searching questions on the understanding and application of around 200 different measures. There is a cumulative score, based on the answers to the questions, with a maximum score of 100. The highest scores achieved to date are around 80. Victoria State plans to use this approach for audit of the Drinking Water Quality Guidelines. The aim, certainly in Melbourne, is to move to the ISO 22,000 standard, which is in the same format as ISO 9001 and ISO 14001, and is described as the tool to harmonise all HACCP based standards. Clearly, this approach leads to highly developed management systems, which should have a big impact on the quality of water supply operations, and thus the safety of drinking water.

The Australians, and others who have made so much progress in this direction, are to be applauded, but I have two concerns. One relates to badges; this section has referred to a number of management standards with success resulting in an organisation being awarded a "badge". In the pursuit of badges, it is important to remember that the primary objective is safe drinking water, not management awards. My other concern, and far more important, is that other countries, which are not yet as well developed, could either attempt "to run down this path" before first learning "how to walk", or could be frightened off by the complexity of the approach and shy away from developing basic water safety plans. I will now consider initial approaches in developing countries.

DRINKING WATER SAFETY PLANS IN DEVELOPING COUNTRIES

Whereas in developed countries, water safety plans are a means of refining operational procedures, in the developing world they also provide a valuable tool for raising the awareness of the fundamental causes of water supply problems, and for establishing investment priorities. A detailed case study [28] was carried out in Uganda. This produced recommendations which included attention to leakage control to reduce ingress into the distribution system, and gave greater emphasis to monitoring chlorine residuals rather than microbiological analysis of water samples. The study showed the importance of giving emphasis to operational management and of the need for management controls from catchment to tap. Good progress has been made on the development of water safety plans in rural areas. In Bangladesh, the approach has been to develop plans according to the type of technology used. The approach has been well accepted by the communities who take responsibility for day-to-day monitoring. Examples for dug wells, tube wells, pond sand filters and rainwater harvesting, are available on the International Training Network Centre, Bangladesh University of Engineering and Technology website [29]. I give below an example of implementation in a developing country, based on a pilot project for an urban water supply, in Ghana.

Ghana water company limited

In Ghana, there is one public utility, Ghana Water Company Limited (GWCL) supplying all urban areas. It is regulated on tariffs and drinking water quality by the Public Utilities Regulatory Commission (PURC). Water resources are the responsibility of the Water Resources Commission (WRC) and the Environment Agency is responsible for monitoring the catchments. In April 2004, there was an initiation workshop, organised by PURC, for a pilot project to develop an approach to drinking water safety planning in Ghana. The Weija Treatment Works supply system, within the Densu Catchment Basin, which supplies part of the capital Accra, was chosen for the pilot project. A project team was formed with members from each of the organisations mentioned above. The team included a member from the Ghana Consumer Association, who was able to provide input on the consumer end of the supply chain, and to stress the importance of an integrated approach to improving the water supply for the benefit of the community. A risk assessment, a review of controls, a review of monitoring requirements, and the establishment of reaction plans and verification measures, were completed by the team in early 2005. There were a number of key recommendations for action, covering each part of the supply chain. On the catchment there was a need to control the amount of nutrients reaching the lake because this was resulting in the formation of algae and difficulties at the treatment plant, with a high likelihood of particle breakthrough. Requirements on the clean-up of the Densu Basin catchment included relocation of informal farming communities, the introduction of good forestry and logging practice, the introduction of better fishing practice, and the control of leachates from waste tips. This will require a long-term programme, and in the shorter-term there is the option of GWCL considering the construction of a floating intake, to allow water to be taken from across an area of the lake, and at depths where there is a lower concentration of algae.

The treatment works was in good condition. Apart from the problems associated with algae removal, there were no aspects requiring significant investment. There were concerns over the unreliability of chlorine supplies and the need for improved monitoring of chlorine dosing.

The situation with the distribution system was very different with the need for replacement or refurbishment of old pipes, the need for replacement of valves and other fittings, and the need for active leakage control. There were no accurate figures on leakage levels but, given the known poor state of the infrastructure, it could have been as high as 50–60%. From the point of view of drinking water safety, the concern about the state of the system related to the inability to maintain 24 hour supplies. Intermittent supplies, with no pressure in parts of the system for many hours, create the risk of ingress of contaminated water from the ground which may contain faecal matter. When water begins to flow again in those pipes, the contaminated water is delivered directly to the point of supply to consumers without any opportunity to make it safe, other than by boiling. This risk received a high score due to the likelihood of daily occurrence with the potential to affect a large population. Investment in the distribution system to reduce leakage would achieve two things; it would provide perhaps an additional 60% water available for consumers and allow for a 24 hour supply to those consumers connected to the distribution system. Such investment would have far greater value than additional resources or an extension to a treatment works.

It is important at this time to mention the road tanker operations which, in effect, extend the distribution system by serving those people living out of reach of the system, and also those who have infrequent piped supplies. Drinking water can be contaminated when the tankers are being filled, within the tankers if contaminated water has been transported, and at the point of delivery to the consumers. This operation was identified as a significant risk to the safety of drinking water. In Accra, the road tanker operators have formed associations to bring some discipline to the tankering operations. PURC, working with the Tankering Associations and GWCL, has produced Tankering Guidelines [30] which include provision for the periodic disinfection of the tankers, and for random samples to be taken and tested by GWCL. The Guidelines can be considered as a useful model for use in other countries.

The pilot project has been successful in achieving an integrated approach in the consideration of risk from catchment to tap, in the development of water safety plans. The pilot has shown that, although the problems are almost certainly known already by those struggling with them day by day, the key requirements for improvement are highlighted by the water safety plan approach, thus giving the priorities for investment. Above all, the approach developed in Ghana concentrates on a few major risks with associated critical controls, and involves minimum documentation. It is an approach suitable for adoption by other developing countries as a first stage. Once the major risks have been addressed, the assessment approach can be repeated and, over time, operational procedures can be developed to match those in existence in the developed world.

The water safety plan approach provides the basis for audit by quality regulators and, in Ghana, PURC has developed regulatory procedures with emphasis on checking the effectiveness of risk management procedures. Sampling and analysis remain necessary for verification and enforcement of standards. The regulatory skill is in achieving a balance between the two parts of the audit. It is important that regulators, in responding to the change of emphasis from end of pipe testing to checking the effectiveness of water safety plans, do not encroach on the management responsibilities of water utilities.

VERIFICATION AND STANDARDS

Returning to the general water safety plan procedure, the final step is verification and, for the water utility, this will normally mean compliance with drinking water quality standards. The forms of numerical standards were covered in an earlier section. Europe, the United States of America, Japan and some other countries have developed their own approaches to the setting of standards; all of these countries take the WHO Guidelines into consideration but also take decisions on standards based on their own assessments. It is not sensible for other countries to consider making their own assessments, because the WHO guideline values themselves are based on the best scientific evidence available at the time, and it requires any other organisation to match that available expertise. Also, now that the WHO Guidelines are web-based with "continuous" updating, they provide the most up-to-date position on any parameter. As an alternative, should another country consider adopting European, American, or any other country's standards? I would say no because compared with the "purity" of WHO guidelines, nationally derived standards can be based on objectives other than healthy drinking water. Some standards have been set in the mistaken belief that drinking water standards can protect the environment. "Putting the horse after the cart" does not work. Drinking water standards should be based on health and aesthetic considerations, and the environment should be protected through a combination of environmental objectives and pollutant controls. The advice therefore is to adopt WHO Guidelines. Should a country have a concern over a substance not yet covered by WHO, it should produce its own interim standard based on available information, and then persuade WHO to take on that parameter for ongoing consideration and inclusion in the Guidelines.

In adopting WHO Guideline parameter values, a country should consider which parameters need to be included and the form and frequency of monitoring. In the European Directive [31] some of the parameters are considered to be mandatory and others are indicator parameters. The distinction is that those listed as mandatory must be complied with, whereas indicator parameters have to be monitored, and action taken only if an exceedance of the parameter value is a risk to health. This has the benefit of focussing investment on meeting those parameters of significance to health. The Directive also provides for two different monitoring regimes, those of check and audit monitoring. Check monitoring, with higher monitoring frequencies than audit monitoring, is required for microbiological, aesthetic and treatment parameters. Ideally, monitoring frequencies should be based on the expected variability over time. In practice it becomes too complicated to consider having an individual frequency for each parameter. It would also be very expensive, as it is much more cost-effective to use the same samples for a number of different analyses. The result is that sampling programmes have to be a compromise between that ideally needed to reflect the differences between parameters and practical considerations related to taking samples and cost.

The locations of sampling are important. It is not sufficient to sample at treatment works because the quality of water can be affected in a number of ways in the distribution system. The microbiological quality should be checked at the treatment works, at any storage points in the distribution, such as service reservoirs and water towers, and at consumers' taps. If there is an exceedance of a standard at any point, it is then possible to determine the source of contamination and make necessary improvements. Other parameters should be checked from samples taken from consumers' taps. One parameter is unique, at least in the way it is treated in the European Directive, and that is nitrite. There are two reasons for this. Firstly it needs to be considered in conjunction with nitrate. Secondly, it can be produced by microbiological activity in those distribution systems supplied by water which has been disinfected by chloramination rather than chlorination [32]. The health concern determining the EU approach to nitrate and nitrite combined, is the risk to bottle-fed babies of methemeoglobinemia (blue baby syndrome). The approach is to set a standard for nitrate of 50 mg/l at the tap, to set a standard for nitrite of 0.5 mg/l at the tap, but also to have a standard of 0.1 mg/l for nitrite at the treatment works. The combination of the two at the tap is controlled by the formula in which (nitrate/50 + nitrite/3) must be equal or less than 1. This formula is taken from the WHO Guidelines in which the guideline value for nitrite is 3 mg/l but, due to uncertainty on the impact of nitrite, the EU has opted for a much more stringent standard for nitrite. A review [33] of the guideline values for nitrate and nitrite was carried out by WHO in 2011. The guideline values have not changed. A brief description of each of the parameters in the EU Directive, as transposed into UK regulations, is given in Part 1 of the Drinking Water Inspectorate's Annual Report of 2005 [34].

It is not the purpose of this chapter to give a comprehensive review of parameters and standards but rather to identify the general issues to be considered in the establishment of standards and regulation. One such policy area is how to protect against *cryptosporidium*, with different approaches being taken in the USA and the UK. *Cryptosporidium parvum*, the form affecting humans and animals, is a single-celled animal (a protozoa), which can survive outside a host for months. The infective stage of the organism, the oocyst, is 3 microns in diameter, and millions of oocysts are excreted by infected humans and animals. They are always present in the environment. *Cryptosporidium* [35] is resistant to disinfection, particularly chlorine, so the best protection is to have processes which will physically remove the oocysts in treatment. In the United States, [36] the regulatory approach is to require specific levels of treatment, measured as numbers of log-removal (each log reduces the numbers of organisms to one tenth as described earlier in this chapter) according to the risks identified determined from two years of monitoring data. Each treatment process qualifies for so many log-removal. In England and Wales, following some serious outbreaks of waterborne cryptosporidiosis, the Government introduced specific regulations (1999) [37] involving a risk assessment, a numerical limit on the numbers of oocysts in a 24 hour continuously taken sample (with a minimum flow through the sampling device of 40 litres per hour), or the option of a treatment process capable of continuously removing particles of less than one micron in size. The limit of less than one oocyst in 10 litres of water was not based on a known infectious dose, but on experience in knowing that outbreaks of cryptosporidiosis have not occurred in "conventional" coagulation and filtration systems that are working normally. Exceedance of the standard was a criminal offence, but water companies had a defence of due diligence. The continuous sampling device was not only an important device for regulatory purposes, but also provides a very useful tool for water utilities to measure and optimise the performance of their treatment plant. The sampling result was much more accurate than so-called "grab" samples, providing for a much higher oocyst recovery. In 2006 following the virtual elimination of *cryptosporidium* incidents Government issued a consultation document, following which it was decided that continuous sampling monitoring was no longer necessary, and that public protection could be achieved through water safety plans. The changes [38] reduced the regulatory burden, which was a welcome step, and reflected the emphasis of a move to a risk management approach in line with the Bonn Charter and the WHO Drinking Water Guidelines.

INTERIM DRINKING WATER QUALITY STANDARDS

In countries where meeting standards based on WHO guideline values will require high investment over many years, it makes sense to set interim standards, with emphasis on meeting the microbiological standards. This has two benefits: it provides the means of focussing investment over the next planning period, and also provides water utility managers with achievable targets. It can be very disheartening to know that the required standards are out of reach, and can result in a lack of motivation for improvement. On the other hand, achieving interim standards can give a sense of achievement and the motivation to achieve more. Once interim standards have been met the "bar can be raised" to aim for the next level of improved quality. Of course, setting interim standards requires care to ensure that parameters occurring at levels of immediate threat to health are included, with standards set which are achievable with available and affordable technology. It is important at the outset to have carried out an extensive survey of water source quality to identify which parameters should be included in the interim standards. It is not possible to be prescriptive on this, but in many cases it can be expected that there will be great benefit if the first level of interim standards are based wholly on microbiological considerations. Such interim standards should include chlorine residual measurements in the distribution system. However, interim standards should be a means to an end, and that end should be the eventual achievement of standards based on WHO guidelines.

A WORD ABOUT DISINFECTION

Disinfection, especially with the use of chlorine, is an emotive subject. The general public has an understandable concern over the use of a "dangerous" chemical – after all chlorine gas is used as a "chemical weapon". This concern is exacerbated by scare stories about disinfectant by-products being carcinogens. So what is the true situation? Yes, chlorine is a toxic gas, and great care has to be taken in its handling. The main risk is in handling it at a water treatment works, rather than in its use in disinfecting water. In fact, disinfection of drinking water with chlorine has probably saved more lives than all medication ever administered. There was a dramatic reduction in deaths from waterborne disease following the introduction of a combination of water filtration and disinfection in Europe and North America in the 19th century [39].

The concern over disinfection is due to the formation of undesirable substances which result from the reaction of chlorine with naturally occurring organic substances in the source water. Drinking water quality standards give limits to the levels of such substances in drinking water, and water treatment processes are designed to remove the organic content ahead of the chlorination process. However, very low levels of the undesirable chemicals remain. There have been extensive studies [40] over the last 30 years to try to measure any adverse impact on health. None has been conclusive, but if there is an effect, it is very small.

There are different views held by water consumers around the world about chlorine and its effect on the taste of water. In some areas of the United States of America, consumers feel uncomfortable if they cannot sense the presence of chlorine, because to them it is an indication of the safety of water. In some parts of Europe any sense of chlorine is unacceptable, and water utilities try to minimise the levels of chlorine residual in the distribution systems. In the Netherlands, in those systems that deliver only "dune" water, which contains no substances on which bacteria can grow in the distribution systems, there is no disinfection of the water supply. This is highly desirable, but unfortunately there are few distribution systems in the world where some means of controlling bacteriological contamination is not needed. Incidentally, any chlorine present in water arriving at consumers' taps will dissipate if left in a clean jug for a couple of hours. If kept in a refrigerator, the taste will also be improved, with people then being unable to distinguish between tap and bottled water in "blind tasting" tests. It is worth noting that the price of tap water is much lower, often only around one thousandth of that of bottled water. If people prefer to buy bottled water, and can afford to do so, then it is a matter of consumer choice, but in most developed countries, no one needs to buy bottled water for safety and health reasons. It must be our aim to extend that desirable situation throughout the world.

Whilst on the subject of aesthetic acceptability of tap water, a word about home-filter systems. Often they are used because people want to remove chlorine, or because they do not like "hard" water. The filters generally reduce hardness, which can mean removing desirable substances like calcium and magnesium [41]. The importance of magnesium is discussed in the next section. They can increase the level of sodium in the water which is undesirable and they can make the water highly acidic, so the advice to consumers is "make sure you know what the devices do to your tap water before using them". Remember the tip about putting a clean jug of tap water in a refrigerator for a couple of hours, to give a pleasant cool drink.

Returning to disinfection, there are alternatives to chlorine, the most common being ozone and ultra-violet light (uv), but neither of these provides ongoing control of bacteria in the distribution system. Ozone also produces disinfection by-products, and requires equivalent care in removing substances in the raw water to avoid their formation. The use of uv avoids such problems, but is not very effective if the water is "cloudy", having high turbidity.

Chlorination is likely to remain the most common disinfectant to achieve safe tap water in the developing world. Any concerns over by-products are tiny compared with the benefits in the reduction of waterborne disease. An important lesson [42] on this occurred in Peru in 1991–2. Following the publication of a USEPA report, there was concern over the formation of disinfection by-products, and the Government in Peru decided to stop disinfection of water supplies. In early 1991, there were the first reports of cholera, and by the end of 1992 there had been over 730,000 cases and 6323 deaths. The final death toll was estimated to be 11,000 deaths in the region. This was an over-reaction to an unproven possibility of a health effect from disinfectant by-products, which put a whole population at risk from a waterborne killer. This is why WHO advises that concern over chemical risks should never compromise the priority of protection from microbial pathogens through disinfection [7]. Also recently, WHO has published a report [13] on protecting drinking water in households with reference to the importance of disinfection.

THE IMPORTANCE OF MAGNESIUM TO HEALTH

WHO set up an expert team to review studies on hard water, calcium and magnesium to assess the evidence of beneficial effects regarding heart disease and strokes. The report [43] concluded that epidemiological evidence for the water hardness-cardiovascular hypothesis is still not proven. However, the balance of epidemiological evidence does support the (beneficial) link between magnesium and cardiovascular mortality. One of the studies [44] carried out in Taiwan over the period 1989–1993 showed a significant lower level of stroke deaths in those areas with higher magnesium levels in the drinking water. This understanding of the beneficial effects from magnesium in drinking water raises some important questions for regulation. Where minerals have to be added into water, for example in desalinated sea water or re-use water treated by reverse osmosis, should minimum magnesium levels be specified? Should devices, such as those home-filters which remove magnesium be regulated, or at least should there be warnings to householders that their use could be detrimental to health? My advice to householders is never to use water filters unless it is known what they do to the water. Suppliers have a moral responsibility to provide that information.

MONITORING, AUDIT AND ENFORCEMENT

The end goal of WHO Guideline values will not be achieved without effective monitoring, audit and enforcement. The role of water safety plans in structuring operational improvements in an integrated way has been covered above, and in considering the process of monitoring, audit and enforcement, I do not make any distinction between compliance with standards and verification of operational measures. The same applies to interim standards, with an equivalent need for an audit process.

The monitoring process requires the collection of samples from treatment works, from local storage units, and from the distribution system, with preferably most monitoring being at the point of delivery, whether this is at consumers' taps or standpipes. Ideally, samples should be taken from consumers' premises chosen on a statistically random basis. However, often access to taps in consumers' premises is difficult, due either to cultural restrictions or to consumers' absence during normal working hours. In such cases, it is necessary to take samples from public buildings, such as government offices and schools, located around the distribution system, to be representative of water delivered throughout each system. Care needs to be taken in sampling, and in the transport and storage of samples, to ensure that samples do not become contaminated, and that the state of the samples on analysis is consistent with the condition of the water at the time they were taken. Analysis methods may be specified in regulations, or rely on water utilities to use suitable methods. The laboratories involved should provide sound training for their analysts and operate an AQC (Analytical Quality Control) scheme to ensure that the results of analysis are accurate to the limits specified in regulations. This may sound obvious, but management should give it continuous attention to maintain the standards necessary to achieve effective monitoring. The whole system of sampling and analysis should be overseen either by third party accreditation or by a regulatory agency.

As I said at the outset, there is little point in monitoring for compliance with standards unless there is a means of enforcement action to require the necessary improvements. There need to be legal sanctions to ensure that standards are met, but it is to be hoped that these will not be used. Voluntary and timely action by water utilities is the most effective way. The knowledge that, without voluntary action, sanctions will follow, provides utilities with both the incentive to make the improvements, and the means of convincing their "political" masters, of the need for investment. I believe that the system of enforcement operating in England and Wales is a good model, and one that has delivered virtually 100% compliance with the standards. This leads me naturally to the example of the Drinking Water Inspectorate, based in London.

THE EXAMPLE OF THE DRINKING WATER INSPECTORATE IN LONDON

The Drinking Water Inspectorate (DWI), one of the three water regulators of water services covering England and Wales, was set up at the time of privatisation of the Regional Water Authorities. In Chapter 9, there is a full description of the regulatory system in England and Wales, including the regulation of drinking water quality. The means of establishing investment programmes is part of the enforcement/improvement process covered in the section on the periodic process.

In summary the original enforcement process had the following steps:

- DWI wrote to a water company advising that it was minded to take enforcement action unless the company offered a legally-binding undertaking to make the necessary improvements.
- The company would respond offering an undertaking "to take some action" to meet the required standard by a particular date.
- DWI, if it was satisfied that the undertaking would achieve the objective, would accept it formally.

DWI would monitor undertakings and should any not be met due to circumstances within the control of the water company, the legal enforcement process would be started. In practice, failures to meet an undertaking occurred rarely, and the undertaking process proved to be an efficient process, resulting in compliance with standards in a much shorter time than that which would be needed with a process involving court procedures.

With a requirement to meet new standards, as was the case in the period starting in the year 1990, the economic regulator, Ofwat, would automatically allow for the additional costs in a price review.

Figures 9.3 and 9.4 in Chapter 9 show the impact of enforcement and investment over the period 1992–2003, with compliance approaching 100%.

It should be noted that due to a European ruling, the undertaking process described above, although recognised as highly effective, has now been replaced by a system of accepting "departures", with compliance required within a maximum period of three years. There are similar drinking water inspectorates in Scotland and Northern Ireland, regulating public sector operations. The UK model is a sound one, and can be adapted for use elsewhere in the world.

REFERENCES

1. Johnston S. (2006). The Ghost Map. Riverhead Books. Published by the Penguin Group, ISBN: 9781594482694.
2. Mortality and Burden of Disease from Water and Sanitation (2004). Global Health Observatory. WHO, 2004.
3. Parasites – Giardia (2012). Centers for Disease Control and Prevention. July 2012.
4. Kappus K. D., Lundgren R. G., Jr., Juranek D. D., Roberts J. M. and Spencer H. C. (1994). Intestinal parasitism in the United States: update on a continuing problem. *Am. J. Trop. Med. Hyg.*, **50**(6), 705–713.
5. Global Burden of Disease Study (2010). GBD10. www.healthmetricsandevaluation.org (accessed July 2013)
6. The Bonn Charter for Safe Drinking Water (2004). Published by the International Water Association.

7. Guidelines for Drinking Water Quality (2004). World Health Organization 2004. ISBN: 9789241546386.
8. Hutton G. and Haller L. (2004). Costs and Benefits of Water and Sanitation Improvements at the Global Level (Evaluation of the). World Health Organization 2004. WHOSDEWSH04.04.
9. Water Safety Plans: Managing Drinking Water Quality from Catchment to Consumer (2005). WHO/SDE/WSH/05.06. World Health Organization 2005.
10. Water Safety Portal. WHO and International Water Association. www.wsportal.org (accessed July 2013)
11. Water Safety Planning for Small Community Water Supplies: Step by step risk management guidance for drinking water supplies in small communities. (2012). World Health Organisation 2012, ISBN: 9789241548427.
12. Hazard and Risk. (2013). Index of Educational Resources. Centre for Occupational and Environmental Health, UK. http://www.medicine.manchester.ac.uk/oeh/undergraduate/onlineresources/hazardandrisk/ (accessed July 2013)
13. Combating waterborne disease at the household level. WHO, ISBN: 9789241595223.
14. Toolkit for Monitoring and Evaluating Household Water Treatment and Safe Storage Programmes. (2012). WHO/UNICEF 2012. ISBN: 9789241504621.
15. Preventing Cryptosporidiosis: the Need for Safe Drinking Water (2013). World Health Organisation, 2013. 2013–91:238, doi: www.who.int/bulletin/volumes/91/4/13-11 9990/ (accessed July 2013)
16. Breach R. and Williams T. (2006). The Pivotal Role of Water Safety Plans, Bonn Principle Series. IWA 2006.
17. Blue Drop Requirements (2011). Department of Water Affairs. Republic of South Africa. http://www.dwaf.gov.za/dir_ws/DWQR/ (accessed July 2013)
18. Hodgson K. and Manus L. (2006). A drinking water quality framework for South Africa: ISSN 0378-4738. *Water SA*, **32**(5). http://www.ajol.info/index.php/wsa/article/view/47853 (accessed July 2013)
19. Policy Issues and Institutional Arrangements: Water Safety Plans for Rural Water Supply in India. Water and Sanitation Programme Report. November 2010. http://wsp.org/sites/wsp.org/files/publications/Water-Safety-Plans-rural-water-supply-India.pdf (accessed July 2013)
20. Drinking Water Safety: Guidance to Water and Health Professionals. (2009). Published by Drinking Water Inspectorate for England and Wales. ISBN: 9781905852406.
21. The Scottish Water (Objectives for 1 April 2010 to 31 March 2015) Directions 2009. www.scotland.gov.uk/Resource/Doc/917/0088613.pdf (accessed July 2013)
22. Water and Health in Hungary: Report of Second Meeting of the Parties to the Protocol on Water and Health. Joint Secretariat to Protocol on Water and Health. http://www.unece.org/env/water/whmop2.html (accessed July 2013)
23. Drinking Water Safety Plannning: A Practical Guide for Pacific Island Countries. WHO/SOPAC Joint Contribution Report. http://www.pacificwater.org/userfiles/file/water%20publication/Pacific%20Drinking%20Water%20Safety%20Planning%20Guidelines%20WHO%20SOPAC.pdf (accessed July 2013)
24. Mullenger J., Ryan G. and Hearn J. (2002). A Water Authority's experience with HACCP. *Water Supply*, **2**(5–6), 149–155. IWA Publishing 2002.

25. Miller R., Whitehill B. and Deere D. (2005). A national approach to risk assessment for drinking water catchments in Australia. *Water Supply*, **5**(2), 123–134.
26. Australian Drinking Water Guidelines (2006). Ref. EH19, EH39. National Health and Medical Research Council Publications, Australia.
27. Stevens M. (2006). Water Safety Plans in Australia. Presentation at WHO Workshop, IWA Congress, Beijing, September 2006.
28. Godfrey S., Nivagaba C., Howard G. and Tibatemwa S. Water Safety Plans for Developing Countries. A Case Study from Kampala. Uganda. Watermark website: http://wedc.lboro.ac.uk/resources/conference/30/Tibatemwa.pdf (accessed July 2013)
29. Examples of water safety plans in rural areas in Bangladesh. Website: http://www.buet.ac.bd/itn/pages/apsudocs/WSP_full_document.pdf (accessed July 2013)
30. Water Tanker Service Guidelines for the Utility and Operators of Tanker Services. (2007). Public Utilities Regulatory Commission, Ghana 2007.
31. Drinking Water Directive (98/83/EC) http://eur-lex.europa.eu/LexUriServ/LexUriServ.do?uri=OJ:L:1998:330:0032:0054:EN:PDF (accessed July 2013)
32. Regan J., Harrington G. and Noguera D. (2002). Ammonia and nitrite-oxidising bacterial communities in a pilot-scale chloraminated drinking water distribution system. *Applied and Environmental Microbiology*, January 2002, 73–81.
33. Nitrate and nitrite in drinking water: Background document for development of WHO Guidelines for Drinking-water Quality. (2011). WHO/SDE/WSH/07.01/16/Rev/1. World Health Organisation 2011.
34. Drinking Water (2005). A Report by the Chief Inspector, Drinking Water Inspectorate. London, June 2006.
35. The Third Report of the Expert Group on Cryptosporidium in Water Supplies. (1998). Drinking Water Inspectorate, London.
36. Long-Term 2 Enhanced Surface Water Treatment Rule. (2006). US Environmental Protection Agency, Washington DC, January 2006.
37. The Water Supply (Water Quality) (Amendment) Regulations (1999). Drinking Water 2000. A Report by the Chief Inspector, Drinking Water Inspectorate. London, June 2001.
38. The Water Supply (Water Quality) (England and Wales) Regulations (2000). SI No.3184 (Unofficial Consolidated version 2010) (PDF 896KB) Website: http://dwi.defra.gov.uk/stakeholders/legislation/wqregs2007cons.pdf (accessed July 2013)
39. Governance for Sustainable Human Development (1997). A UNDP Policy Document. United Nations Development Programme, January 1997.
40. O'Neill G., Mole N., Horth H. and Fielding M. (1994). Disinfectant By-products: Occurrence and Significance to Health. Report No FR0481 Foundation for Water Research, November 1994.
41. Report on WHO Meeting of Experts on the Possible Protective Effect of Hard Water Against Cardiovascular Disease, held in Washington DC, USA April 2006. Public Health and Environment WHO Geneva 2006.
42. Chlorination M. X. and the Law of Unintended Consequences (August 2003). Kimballs Biology. http://users.rcn.com/jkimball.ma.ultranet/BiologyPages/C/Chlorination.html (accessed July 2013)

43. Calcium and Magnesium in Drinking Water: Public Health Significance (2009). World Health Organisation 2009. ISBN: 9789241563550.
44. Yang Chun-Yuh Calcium and Magnesium in Drinking Water and Risk of Death From Cerebrovascular Disease. http://stroke.ahajournals.org/content/29/2/411.full (accessed July 2013)

Chapter 5

Transparency and public participation

It's good to talk

Transparency is often seen in the context of dealing with corruption. I am sure that it is no accident that the international organisation dedicated to combating corruption is called Transparency International. However transparency is also important for many other reasons because the lack of it allows politically unpalatable information to be hidden. Some examples are:

- A municipality using water revenue for other purposes, with a resultant neglect of the water infrastructure;
- A local politician pressurising a water service corporation to hide a water quality incident;
- A local authority not wanting to admit that a service contract was poorly specified and poorly monitored;
- A private utility failing to carry out required and funded improvement programmes.

Public participation goes hand in hand with transparency because how else can citizens have confidence that they are receiving accurate and honest information? Generally, in the developed world, citizens are no longer willing to take things on trust; they want to be assured that their tap water is safe to drink, and they wish to have some say in water and wastewater developments that affect their lives and their environment. From my observations, for most people, it is more important that they have the opportunity to be involved, than the involvement itself. Unfortunately, this means that those who do take the opportunity and become involved are not necessarily representative of the majority of the population. This raises the question of how to obtain a balanced involvement, and avoid policies being made and actions being taken that suit those with specific narrow interests, to the possible detriment of the majority.

In the developing world the situation is very different, with a need for the underserved to have some means of improving their situation. It is argued that civil society [1] needs to be involved to hold regulators to account, especially in the global south where they may not have the required level of independence from government. NGOs have an important role to play; for example Water Aid whose citizen action initiative is discussed later in this chapter. Water consumers seek involvement through local community associations which lobby government for improvements to their water services. There are good examples of this in India and Africa, some of which are described later in this chapter. Local community associations can provide great benefit for their governments and water service providers, as they are able to give important input on problems and solutions. In particular, they can advise on approaches to payment, such as stressing the need for easy instalments for access charges. Such associations, with focused objectives to improve the local services, avoid community action becoming associated with wider political movements, as has happened in Bolivia.

Table 5.1 Table based on Arnstein's ladder with brief descriptions of each rung.

Level	**Form of participation**	**Description of participation**
8	Citizen Control	Examples given by Arnstein are funding of communities to run their own co-operatives and development projects
7	Delegated Power	Citizens have dominant decision-making authority on a specific item or plan, through a majority of seats
6	Partnership	Power is redistributed through negotiation between citizens and power holders e.g. joint policy boards
5	Placation	A token degree of influence by having hand-picked individuals on Public Bodies
4	Consultation	A further step (Public meetings) but offers no assurance that concerns or ideas will be taken into account
3	Informing	An important step but tends to be one way with no channel for feedback
2	Therapy	Citizens who may be complaining are given therapy to divert their attention – not relevant to water but healthcare
1	Manipulation	Citizens put on advisory committees for the express purpose of educating them or engineering their support

Source: Author, taken from information in [2].

LEVELS OF PUBLIC PARTICIPATION

Levels of public participation were very effectively defined in 1969 by the Arnstein ladder (Table 5.1) [2]. There have been many variations of the ladder but, by and large, Shelley Arnstein's structure has remained a useful classification. Over 40 years later, almost any paper on public participation includes reference to Arnstein's ladder. Humphreys [3] refers to "its continuing usefulness and relevance". In the developed world the two bottom rungs of the ladder tend now to be discarded, as it is generally felt that governments are now serious about public participation. Arnstein had labelled levels 1 and 2 as nonparticipation (Figure 5.1). The question now in the developed world is the extent to which citizens are given power over local plans. In the developing world, with less experience of public participation, it is important that governments avoid the temptation to manipulate citizens by issuing incomplete and biased information. Real and effective public participation may be seen by public officials as diminishing their power, rather than a means of strengthening their authority achieved through gains in public support and in information on citizens' requirements. The discussion on sanitation policy in Chapter 1 highlights the need for participation across the various domains in a city.

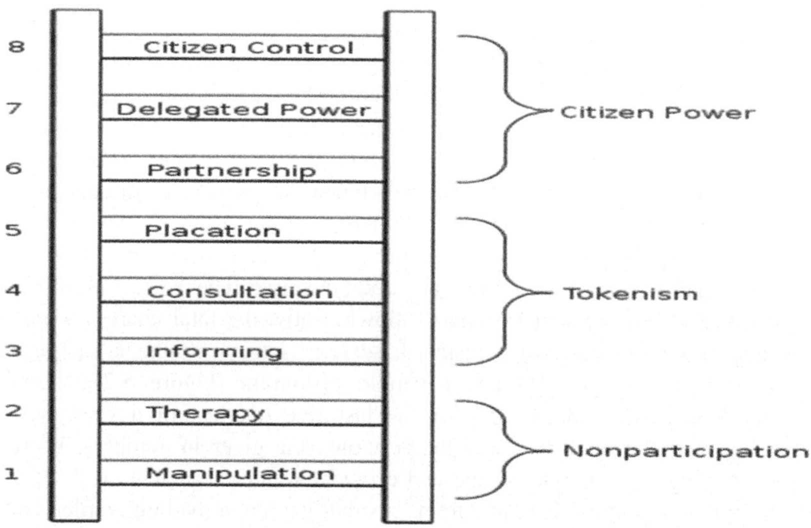

Figure 5.1 Arnstein's ladder (*Source*: Shelley Arnstein [2]).

A much simpler framework of three levels – consultation, presence and influence is given by Goetz and Gaventa [4]. OECD [5] also give three categories, those of information, consultation and active participation. Another version was developed by Tameside Council [6] in the UK, with three levels of increasing involvement, from informing, through consulting to involving/devolving; a table in the paper lists 26 different items of consultation with the top level of involvement including service-user representation on decision making bodies and neighbourhood appraisal. Although related to general local authority functions rather than to water services, it does give a good illustration of increasing engagement. Others will have different views on the position of some of the items, as it depends very much on the local political climate. For example, I find it difficult to position referenda within the ladder. In one sense, a referendum, as shown half way up the ladder, can be merely consultation but, to my mind, it can be much stronger than the normal level of consultation, as the public can be given a veto on proposals, allowing them a lot of influence. The veto form of influence is different from the Arnstein definitions of placation or partnership. This example illustrates the limitation of trying to develop a one-dimensional scale of influence, but nevertheless I see it as a valuable comparative tool.

LEVELS OF PARTICIPATION IN WATER SERVICES

It is instructive to consider how various common water service forms of communication or participation relate to the Arnstein ladder, and whether they can be regarded as informing, consulting or involving the public. Some examples are given in Table 5.2. It is important not to decry those forms of participation with lower "scores" as, in situations in which there is a good water service, and there is trust in the water service provider, consumers can be satisfied by receiving good information without the need for any form of direct control. Even the water bill can be developed to provide information in a clear and attractive way. The form of bill issued by Yarra Valley Water in Melbourne, Australia, which won a Communication Award at the International Water Association Congress in 2006, is a good example. The Yarra Valley Water bill provides a clear account summary, showing how the total charge is broken down into water usage, sewage disposal, service charges, drainage charge and an annual parks charge. It gives a simple histogram (Figure 5.2), showing how water usage has compared over the last five quarters of a year, with a comparison with the same period in the previous year given in numbers. There is a chart showing typical water usage and efficient water usage with 1, 2, 3, 4, 5 and 6 people households with no garden, a small garden, a medium garden and a large garden, so that each household can see how it relates on water usage efficiency. Importantly, it states that the usage figures are updated for each bill issue, based on real consumer data. The second sheet of the bill provides a more detailed explanation on the bill calculation, and gives water saving tips and ideas. The bill is easy to understand, informative and very persuasive on water conservation.

Table 5.2 How forms of participation score against the Arnstein ladder.

Form of participation	Description of participation	Arnstein rating	Informing consulting involving
Water Bills	These can be either be just a demand for payment or a source of information	0 or 3	Inf
Information leaflets	Printed information available at water service organisation offices or elsewhere such as in libraries	3	Inf
Website information	Can be either passive as with printed information, or can be more active if there is a means of making comments or seeking explanations	3 or 3+	Inf
Media briefing	Delivery of information, such as an Annual Report with opportunity for questions and discussions	3	Inf
Consultation papers	These might be issued by government as a serious approach to consultation by raising questions or issues for response	4	Cons
Public Meetings	Typically, these are held when a proposed project would affect local people such as the construction of a reservoir	3 or 4	Cons
Water Consumer Councils	If independent, can give some real influence to consumers especially in relation to consultation and compensation	4 or 5	Cons or Inv
Local Community Groups	Typically these come into existence in communities where there is a poor or non-existent service. Their influence is dependent on the political climate	4–6 Can develop into 7–8	Seeking Inv
Referenda	Rarely used on water matters, but in Switzerland the public can veto proposed developments	4+	Inv
A Public Role in Regulation	Public representation on regulatory bodies	5	Inv
Local Action Agenda	Situations in which local people form a local organisation to develop and operate their own water service	7–8	Inv

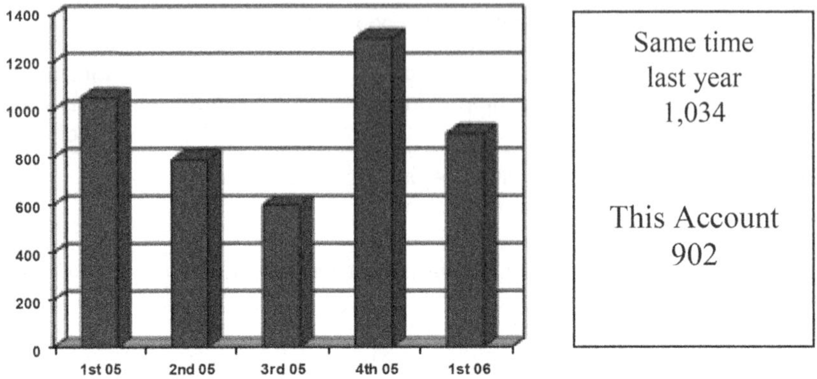

Figure 5.2 Part of a Water bill giving usage pattern (*Source*: Yarra Valley Water, Australia).

PUBLIC PARTICIPATION AS PART OF ACCOUNTABILITY

As demonstrated in case studies, some of which are given in this and later chapters, effective public participation is a key element in successful water service provision. It can be argued that public participation takes place through elected politicians and policy makers but the process of communication is very slow and generally not very effective. This is called the long route of accountability [7]. What is required is direct public access to water service providers (Figure 5.3).

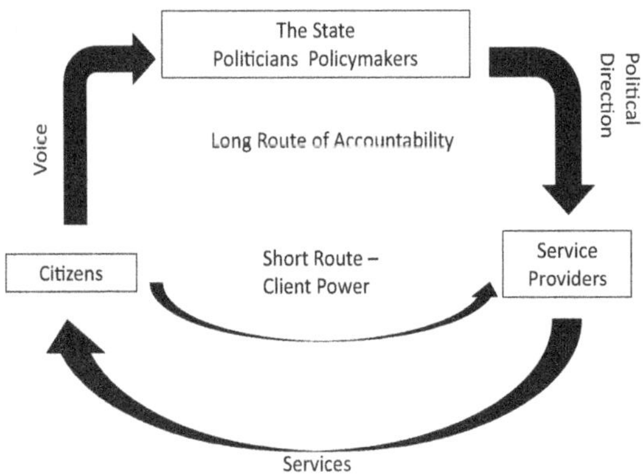

Figure 5.3 Long and short routes of accountability (*Source*: World Bank [7]).

The World Bank includes public participation in its tools of accountability [8]. Similar categories of involvement to those examples given above are used (Table 5.3). The three categories are given as information, consultation and participation. There is also a fourth category, that of Recourse/Redress, the mechanisms of which are important not only in addressing failures in meeting defined service standards, but also as part of transparency.

Table 5.3 Participation and accountability.

Information	Community outreach and ad-hoc meetings Publication of performance data On-demand information provision
Consultation	Forecast surveys Retrospective performance and perception surveys Structured consultation processes Membership on advisory bodies
Participation	Membership on decision making bodies Involvement in the execution of specific utility activities Participatory budgeting Ownership of utility
Redress/Recourse	Utility complaint mechanisms Third party complaint mechanisms Legal recourse and redress

Source: World Bank [8].

A number of tools are used to stimulate accountability. Some [9] are listed below

- *Participatory Budgeting*: a process through which citizens participate directly in the different phases of budget formulation, decision-making, and monitoring of budget execution.
- *Independent Budget Analysis*: citizens study a public budget to build their understanding, lobby for adjustments, and monitor expenditure.
- *Public Expenditure Tracking Survey (PETS)*: a quantitative survey of the supply side of public services.
- *Citizen Report Cards (CRCs)*: surveys that solicit user feedback on the performance of public services.
- *The Community Score Card (esc)*: a community level tool that links service providers to the community and facilitates assessment of the services towards negotiated improvements.
- *Social Audit*: citizens engage in collecting impartial and evidence-based information, which is used to expose and deter corruption and mismanagement.
- *Citizens' Juries*: selected members of a community make recommendations to decision-makers on complex issues after a period of investigation.
- *Community Radio*: a non-profit service provides an interactive medium of communication that offers opportunities for listeners to participate. It deals with local issues in local languages and in a cultural context.

- *Enumeration*: A process used mostly by residents of urban informal settlements to establish the numbers of households and the quality and quantity of facilities available. Information is used as an advocacy and negotiation tool.

The citizen report card (CRC) approach [10] was pioneered by the Public Services Centre in Bangalore, India. The process involved carefully prepared design of questionnaires and structured samples of citizens carried out by a market research organisation. It assessed the levels of satisfaction of citizens with water, power, municipal, transport, housing, telephone, banks and hospital service agencies. The first CRC carried out in 1994 showed public satisfaction to be low with corruption as a major problem and a lack of customer orientation and service. The process was repeated in 1999 and in 2003. Following the results of the first CRC the State Government and the service agencies embarked on a series of reforms. The greater transparency and responsiveness to citizens' requirements was reflected in significant higher scores in the subsequent CRCs.

Kenya embarked on a water service reform programme in 1999. The reforms were implemented through the Water Act 2002. The main features of the reform were separation of policy and delivery, decentralisation and participation. The regulatory/management structure arising from the reform of 2002 is shown in Figure 5.4.

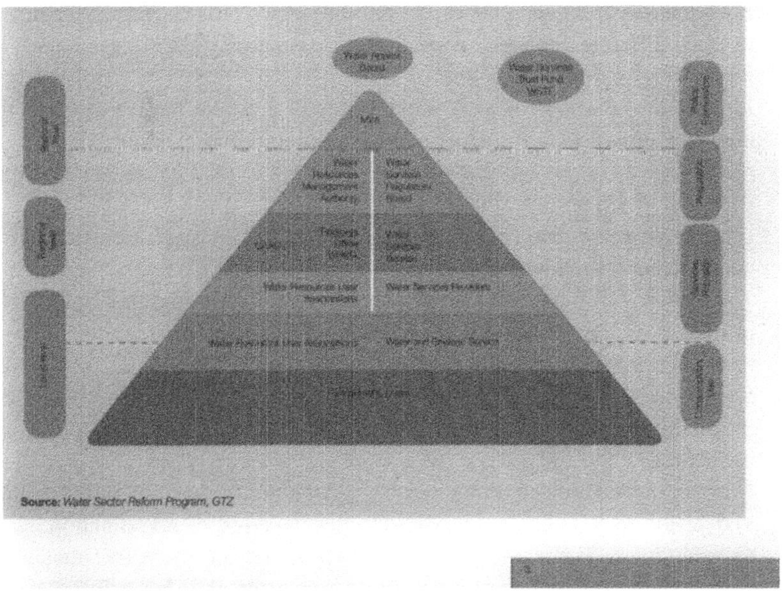

Figure 5.4 Structure of reform 2002 in Kenya [11].

One of the water reforms' stated goals was to increase stakeholder and beneficiary involvement. It became evident that this had not been addressed, with the public not understanding the reasons for the reforms or who was responsible for what. In response to criticisms the water service providers considered a number of public participatory tools and based on the good experience in Bangladesh chose CRCs [11]. The Ministry of Water and Irrigation endorsed the proposals and CRCs were introduced in the three main urban centres, Nairobi, Mombasa and Kisumu. Implementation was by way of consortia comprising government, water service providers and civil society. Social society acting as secretariats led the initiatives.

The stated objectives of the CRCs were to:

- Create larger scale and more direct links between citizens, water and sanitation policy makers and service providers
- Deepen citizen understanding of the water and sanitation reforms, their rights and of the collective service delivery issues faced in the three cities in Kenya
- Use the evidence from the CRCs to support citizens to carry out informed dialogue and advocate for improved services for both the urban poor and non-poor

The water service providers were willing to respond to the feedback from the public and made significant service improvements during the CRC action period. Also, the citizens were able to appeal to government on governance matters, and in Mombasa for example the Minister made appointments to a water services board to reflect public interest. The Water Services Regulatory Board acted on the CRC feedback and ensured that there were mechanisms for resolving issues and complaints. Water Action Groups made up of citizens from resident associations were set up to monitor water and sanitation provision and to follow up on unresolved complaints.

The media were influential in reporting the findings from the CRCs. The CRCs provided good information on citizens' priorities for improvement, which were:

- A more reliable water supply and improved reliability were more important than cost reduction (Author comment: This reinforces the argument that citizens are willing to pay for a good service)
- Citizens using kiosks want their sources to be closer
- Citizens want access to more public toilets and greater access to the sewer network
- Citizens want more direct access to service providers, firstly face to face, and secondly through local radio

The main conclusions arising from the Kenyan experience were considered to be:

- A partnership approach is beneficial with service providers willing to act on findings

- Partnerships take time to develop, requiring investment to build skills, to clarify objectives and reach agreement on processes and outputs
- The importance of evidence-based information tools such as CRC, and I would add
- The value of having an independent regulator to oversee the action processes

CRCs are an important part of the WaterAid Citizens Action approach which is described later in the chapter under The Role of Governments.

MEDIA BRIEFING

Often the media are treated as if they are the enemy, and to be avoided if at all possible. Certainly, modern reporting, whether it be national newspapers, national radio or television, does seems to demand bad news, so there is a tendency for reporters, or often their editors, to give bad news the headlines. Good news is either ignored or it appears as a small column on an inside page. In my own experience, the situation with local media, and local radio in particular, is different, with a much greater willingness to communicate facts to the public. In this context it is interesting to note the value which Kenyan citizens place on local radio communication. During an emergency, local media become a vital part of communication between the authorities and the public. This does not mean that there are not well-meaning national reporters, but simply that their editors tend to cut good news stories. It is therefore necessary to provide reporters with information, or a story which they can make into a newsworthy item. For example, when launching a regulator's report on drinking water at a press conference showing that tap water is of high quality, it is important to try to add another dimension to provide a storyline for reporters. In the UK, a comparison with bottled water has proved to be successful, resulting in newspapers running "blind tasting" events comparing tap water with bottled water, and then writing complimentary articles on tap water. It is a question of the media having trust in the authorities. This takes time to be established and requires total honesty, with willingness to be open with information when something has gone wrong. It is difficult for a water service provider to establish a position of complete trust with the media, because there is always the suspicion in the back of a reporter's mind that "they would say that, wouldn't they"! There is great benefit in having independent regulators whose role is to be totally objective and report factually, but even they have to establish trust through transparency and competence. This vital area of regulatory independence is discussed in Chapter 2. The extent to which (or more perhaps when) social networks will replace newspapers, TV and radio as the dominant form of communication is unclear at this time. The role of social networks is discussed later in the chapter.

PUBLIC MEETINGS

Public meetings can be a useful form of communication as they bring citizens in direct contact with government or others in authority. However, they are often difficult in practice because there is a tendency for them to be held only when there is something controversial to be discussed, such as a proposed new reservoir. In these situations, there is likely to be a strong group opposed to a scheme, and a local protest movement may be under way. This means that a meeting may become confrontational in nature. Even if those in the locality agree with the need for a new development, inevitably there is the NIMBY (not in my back yard) factor. In the UK, public meetings on water matters tend in peoples' minds to be associated with public enquiries following some form of problem or difficulty. One example is the public meetings that followed the Camelford incident [12] in the South West of England in which a relief driver from a chemical supplier, unfamiliar with the treatment plant layout, poured a delivery of aluminium sulphate into the wrong tank, resulting in contaminated water going into supply. Another example is the public meetings held after the 1995 drought in Yorkshire. Thus if governments, or water service providers, begin suddenly to hold public meetings it is likely to be viewed with suspicion that something has gone wrong. The lesson from this must be that if public meetings are chosen as a desirable means of public consultation, they need to be held on a regular basis, quite independent of any specific development, in order to establish a working environment of trust. Local Authorities, who hold their business meetings in public, tend to use public meetings extensively as a means of explaining future plans for a city, or for obtaining public feedback on proposed new developments. IDeA, the UK Improvement and Development Agency for Local Government, has issued some valuable advice [13] on planning and holding public meetings.

WATER CONSUMER COUNCILS

The establishment of Water Consumer Councils is an attempt by governments to provide the public with a real voice on water matters, and to give individual consumers a means of obtaining redress if the water service falls below prescribed standards. They can operate as part of a regulatory body, in which case they can expect to use the regulator's powers to obtain redress in the form of compensation payments. The disadvantage of this is that the regulator may give them only limited independence, with the risk that their input into policy determination may have to mirror the views of the regulator. An alternative model is that they are wholly independent bodies with separate funding, with the freedom to consult fully and openly with the public and to develop and present their own distilled views on policy. In this case, they may have greater difficulty in persuading the regulator that a consumer is due compensation. As with all things, success is not only dependent upon the structure but also relies on effective working relationships.

There has been a long history of consumer councils in the UK beginning with the establishment of Regional Water Authority (RWA) based Consumer Consultative Committees in England and Wales in 1983. The aim at that time was to provide some form of public involvement at the time of the removal of the local authority representation from RWA Boards. These Committees had little authority and did not provide independent consumer representation. On privatisation in 1989, Customer Service Committees were set up under the direction and management of the economic regulator, Ofwat, and later the Water Act 2003 made provision for the replacement of the Customer Service Committees with a wholly independent body, the Consumer Council for Water. During my time working at the Drinking Water Inspectorate I worked closely with the Consumer Council which adopted the name "WaterVoice". My experience was that they provided water service customers with the opportunity for objective input to policy, and they were a strong advocate in dealing with service problems and deficiencies. The successor body, the Consumer Council for Water in England and Wales, has greater independence with the opportunity for a more influential input on government policies. So water consumers in England and Wales have experienced three forms of water service bodies with increasing independence over time. More detail on their roles and policies is given in Chapter 9.

LOCAL COMMUNITY GROUPS AND LOCAL ACTION AGENDA

I take these two together as the term Local Community Group is given both to local groups formed to apply pressure on governments to make changes or improvements, and to local groups who take the initiative in the development and management of their own water services, through local action agenda. Generally, although not exclusively, local action groups tend to be formed in rural areas, and pressure groups are formed in urban areas where it is much more difficult to get access to water sources. Both types of group have an important role to play. I discuss some examples of local community groups and draw some conclusions on the necessary ingredients for success.

La Paz, Bolivia

In the now much maligned concession contract in La Paz in Bolivia, which is described in Chapter 8, there was important input from neighbourhood groups, in a SIDA/WASH/UNDP (ie Swedish International Development Agency, Water, Sanitation and Hygiene Program of the World Bank, and the United Nations Development Program) sponsored pilot project, who were able to assist in the development of a scheme to give access to the poor. In particular, the neighbourhood groups were involved in condominial systems in which local people provided labour to install low cost distribution systems. The neighbourhood groups were able to advise on setting up payment facilities for low-income communities, and to assist when there were difficulties in making new service connections. Importantly, they were neutral facilitators trusted by the communities.

Accra, Ghana

Some people living on the outskirts of the city of Accra in Ghana do not have access to a continuous supply of tap water. In some areas they are wholly reliant on tanker and cart supplies, and in other areas the tap supplies are intermittent. Local Community Groups have formed to seek improvements in their water supply situation. The first example is that of a pressure group formed to obtain political attention to their problem. The Teshie Concerned Citizens Association [14], who said that they had not received a regular tap water supply for 20 years held a press conference at the end of December 2002 with an appeal to the Ministry responsible for water to restore a regular supply. In February 2003 members of the Association took to the streets in Accra to protest at what they called "the failure of Ghana Water Company Ltd. to restore and maintain a regular supply of water to Teshie". The company response was that one of the main reasons for the shortages was the amount of water taken through illegal connections made by residents in the area. The Greater Accra Regional Minister set up a Committee, which included a member from the Association, to investigate the Teshie water supply situation. The Committee, confirming the seriousness of the situation, reported in June 2003. The report identified eight causes of the problems, including the rapid expansion of the area, and made a number of recommendations. One recommendation was the completion of the development of local storage so that more water could be taken during the times when distribution system pressure was high enough to supply the Teshie area. The Association had been successful in obtaining political focus on their problem.

Since that time, the regulator, the Public Utilities Regulatory Commission (PURC), has instigated three pro-poor pilot projects (personal communication) involving the Teshie Group and two other community groups, those of Glefe and Nima. Each community formed a Water Board to manage the systems. Bulk storage was installed in Teshie, standpipes were put in place in Glefe, and a bulk-storage with standpipes system was built in Nima. There were some problems in construction with contractors, some vandalisation of installations, and some pricing difficulties where tanker supplies were required. However, although the results were variable, each area received an improved service and some important lessons were learnt. In particular, the Glefe Water Board was very successful. It planned to extend the service within the local area, and even considered adding sanitation services. PURC plans to scale up the pro-poor programme to other areas.

Savelugu, Ghana

In this example [15], the need for improvements in the water supply to Savelugu, a town in Northern Ghana with a population of 25,000 people, arose from a high level of the incidence of Guinea worm disease. The piped supply to the town had been discontinued in 1985, and between 1985 and 1999, most people had to rely on dug-outs for their water supply, with these waters being the source of the Guinea worm. In 1999, a partnership was established between Ghana Water Company Limited (GWCL), the public sector supplier, and the community of Savelugu. The community formed the Savelugu Water Board (SWB) to manage the local system. The agreement is that GWCL provides a bulk supply of water to SWB. SWB handles the distribution of the water, the collection of water bills and the payment at full bulk water cost price to GWCL. UNICEF financed 70% of the project cost, with support from others. The major costs were the construction of a replacement 4 km pipeline connecting Savelugu to the GWCL system, and the construction of multi-tap points from which residents collect their water. GWCL donated a 20,000 gallon storage tank. GWCL provided technical support and repaired any major pipeline bursts. The improvement in public health was immediate with the number of guinea worm disease cases down from 667 in 1999 to 28 in 2000. This progress was sustained until in July 2011 a celebration [16] was held in the Savelugu Nanton District to mark the total eradication of guinea worm infection in the country. Although there were problems with GWCL unable to supply enough water, and funding needed to add some groundwater sources, it had been possible to fix tariffs to provide a subsidy for the poor. Distribution system leakage levels were low, with a figure of 15% reported for the year 2002. The key elements associated with success in this case were external funding of the one-off construction costs, good technical support and very importantly, a Water Board committee comprised of both men and women, taking on the management responsibilities for the community. In 2012 it was reported [17] that a 70 km pipeline project funded by The Millenium Development Authority of Ghana, to bring water from a major source to supply the city of Tamale and surrounding districts, had been completed. This included the Savelugu Nanton District, to provide a more reliable water source.

India – Alwar District

There are many good examples of effective public participation on a large scale in India, often in rainwater harvesting schemes, in which local action is taken to develop new systems or to rehabilitate systems which had been in operation previously for hundreds of years. One well-documented example ([18], [19] and [20]) is in the north Rajasthan district of Alwar which since 1985 has been transformed through local community effort, supported by the NGO Tarun Bharat Sangh (TBS). Following a severe drought in 1985, a local community leader suggested to TBS that the poor water resource situation could be alleviated through a traditional rainwater harvesting approach. This involved the construction and rehabilitation of earthen check-dams, which in India have various names, such as johads (Figure 5.5), khadins and ahar pynes, depending on the topography, the type of soil and the details of construction. The check-dams hold back water during the rains so that groundwater is recharged, or the soil is irrigated for agriculture. The work started in a small way in pioneering villages, but grew rapidly when neighbouring villages observed the benefits of the johads. Since 1985 more than 4500 johads have been built in over 1000 villages, raising the level of the water table in the area. As a result the Arvari River, which had been dry for many years, returned to full flow in 1996. This has transformed the economy of the area with growth of profitable agriculture and fishing. The construction of johads was carried out largely with local labour but with some hired skilled labour. Also some materials not available in the village had to be purchased. TBS acted as a facilitator and advised on setting up the village assemblies, called Jal Samitis, to achieve strong community involvement in decision making. It also assisted the villages in obtaining external funding, and established the principle that the communities should pay for at least one quarter of the cost. Later this was increased to one third of the cost. Villages established funds to pay for maintenance of the johads with each family being required to contribute 2.5% of its income to the fund, or to provide one day of voluntary effort per month.

Figure 5.5 An example of a johad (check-dam) (*Source*: Centre for Science and Environment, India).

The individual local management of the water resources has developed into a local water parliament. The re-establishment of the river to normal and continuous flow conditions resulted in the return of fish to the river, and the

Rajasthan Government let a contract for commercial fishing in the river. Although the villagers were vegetarian, and not interested in the fish, they saw this move as a possible step by government to take over control of their water resources. So in 1999, the communities formed the Arvari Parliament to manage the Arvari River and its tributaries. The Parliament, which has no legal status, has 142 members who are nominated by the 72 village assemblies. This body has established rules for the use of the rivers, and has a coordinating committee to handle operations and enforce the rules, albeit on a voluntary basis. The rules include those related to extraction for irrigation, land use and ownership, and protection from industry. In practice, the Parliament is working well with effective resolution of disputes, and is safeguarding the water resources through preventing the growth of water-using industries.

The success of the Alwar initiative has stimulated international agencies, such as the United Nations Development Programme, and the Government of Rajasthan to support the Jal Bhagirathi Foundation [21], which was formed in 2001, to establish water conservation measures in the Marwar belt of the State. TBS is providing advice, including that on setting up local committees, the Jal Samitis, to include women and representatives of the weaker parts of the communities. Since 2001 the Foundation has been very active with support to communities in the construction of over 70 storage units serving over 14,000 people. The Foundation signed an agreement in 2003 with UNDP for a project on "vulnerability reduction through community management and control of water in drought-prone areas of the Marwar region". Another agreement was signed with UNICEF for a project on "community empowerment for drought management". It will be interesting to see the extent to which this concerted effort to expand community-managed johad schemes will yield sustainable water supply systems over the next 20 years.

India – Kolhapur

This example involves a scheme introduced in 1980 for the supply of river water to four villages in the Kolhapur District in Maharashtra. This system was the subject in a Water and Sanitation Program Field Note [22]. The policy was for districts to take responsibility for the operation and maintenance for such systems, but in this case the Kolhapur District did not have the financial resources to do so. The Maharashtra Government persuaded the local administration of the four villages to take on the operation and maintenance of the system. In 1981, the villages formed a joint water management organisation, called a Mandal in India, to operate and maintain the system on an informal basis. Initally there were 419 house connections but, through installing additional pipelines and encouraging more people to connect to the system, by 1999 the number had grown to 1723. Those with house connections paid an initial connection charge, an annual water charge and an annual development charge as part of financing system enhancements. Those villagers without a house connection are able to take water free of charge from public standposts, with each village making an annual financial contribution to the Mandal for the maintenance of this public facility. Over the period 1981–1999, the Mandal was able to achieve a revenue surplus. The accounts are audited each year and are available for public scrutiny. The system is not without its problems, with deterioration of the river water quality and the need for replacement of some of the expensive equipment, such as the pumps, but its overall performance greatly exceeds that of neighbouring systems managed by the District. For example, because it has been necessary to achieve financial stability, the Mandal introduced a policy of disconnection to ensure that householders pay their bills. This was possible without accusations of penalizing the poor due to the availability of public standposts. However, by 1999 only 16 disconnections had been necessary. In comparison, the other District managed systems have no such policy with a much higher level of non-payment and the risk of unsustainable systems. This example demonstrates that self-management generates a high level of community responsibility, and that generally non-payment is not a case of people not being able to pay for water services, but not wanting to pay for them.

Orangi, Karachi, Pakistan

One of the largest and most-cited examples of successful community participation and action is the Orangi Pilot Project, OPP. Orangi is a township west of Karachi city with a population of over 1 million people. The project, which is described by Arif Hasan in his 2001 Stockholm paper [23] was established in 1980 by an eminent social scientist, Akhtar Hameed Khan, who believed that government programmes were failing because government officials and their professional advisers did not have a sufficiently good understanding of the culture, the sociology, or the economics of low-income communities, especially those in informal settlements. His research led to the identification of four important needs: those of sanitation and housing, employment, health and education. Within that group of important requirements, sanitation was seen by the Organi residents as of the highest priority. Four NGOs were formed with the Orangi Pilot Project – Research and Training Institute (OPP-RTI), covering sanitation, housing, education, research and training.

The approach taken was for OPP-RTI to hold meetings with those residents willing to establish a Lane (Street) Organisation and appoint a Lane Manager. OPP-RTI offered technical assistance to Lanes to build their own sewerage systems. Typically a Lane consists of 20–40 houses. Initially, Lanes with a natural drainage channel would build their latrines and sewer drains, and later this was expanded to include connector sewers to take sewage from other Lanes. There has been criticism of the use of natural drainage channels, but initially this was necessary for affordability reasons and because of the limit of community influence. The longer-term intention was for the Karachi Corporation to convert the channels into box trunk sewers leading to sewage treatment works. By 2001, over 6000 Lanes with a total of over 90,000 houses had built their sewerage systems. Due to choice of technology, self-construction and self-financing, the cost per house was only US$16.5, and therefore affordable.

As a result of the success of OPP, the approach has been replicated in other areas, some through local government initiatives, and some with financial support from outside agencies. Some have succeeded, some have failed. Arif Hassan, in his paper, gave the reasons for the successes as follows:

- a combined technical/social organisation which included members of the community;
- a leader who could establish an effective relationship with local government;
- the existence of a good map of the area;
- patience;
- availability of funding for collector sewers;
- advice and training from OPP-RTI;
- effective monitoring and management through regular weekly project meetings;
- transparency of accounts and involvement of local people on the board of the NGO involved.

He gave the reasons for failures as the lack of these above items, plus acceptance of large sums of donor money leading to financial mismanagement, and subsidising of Lane developments which had resulted in disputes, higher costs and loss of empowerment of the communities.

The challenge for OPP-RTI, with resource limitations, had been how to meet the demand for its expertise and experience. It met these through training other NGOs and running lecture series for communities and elected officials.

An update by OPP-RTI in 2012 [24] showed that further progress had been made in extending the approach to other areas with government commitment and finance for main sewers and sewage treatment. Over 106,000 houses are now covered in Orangi. OPP-RTI advised government on sanitation policy resulting in a joint community/government programme. At the time of the 2012 statement the community had invested $1.26 million in toilets and lane sewers about $12/house, which is lower than the earlier figure quoted of $16.5 per house. The approach has been extended to 438 settlements in Karachi, and to 33 towns and 93 villages in Sindh and Punjab Provinces, with a total of over 2 million poor people covered. Government is investing $8.65 million in main sewers and sewage treatment. This success belies the view that Orangi was unique. The successful wider application of the approach suggests that the factors identified by Arif Hassan in his 2002 analysis were in place. Government support and finance for the higher cost infrastructure is a critical part of the approach without which waterborne systems cannot be environmentally sustainable.

San Luis, Agusan del Sur, The Philippines

The training necessary for communities to manage water and sanitation systems in the area of Agusan del Sur in the Philippines has been provided by an NGO, The Philippine Center for Water and Sanitation, through the establishment of a Provincial Water and Sanitation Center. The physical water supply system was provided by the local government for community management. The Center's role was to build the capacity of municipal level professionals to support the community water and sanitation associations. An example of the success of this scheme has been in the community of Barangay Doña Flavia in the Municipality of San Luis [25], where system coverage has been developed to serve 100% of the population of 2687 people. Importantly, the community receives a constant 24 hour a day supply. The approach taken was for the Center to hold orientation seminars for community organisers and to train trainers who would assist the community teams. The training included both the concepts of sustainable systems, and the more detailed aspects related to hygiene promotion, technology and tariff calculations. It was reported [26] in 2006 that 33 other municipalities in the area of Mindanao had established community managed water and sanitation systems.

Tanzania

In its National Water Policy of July 2002 [27], the Government of Tanzania announced a change in government policy from being "a provider of free water for all", to a facilitator of people's initiatives to improve their water and sanitation services. This was good news in recognition that "free" water services are not sustainable (see Chapter 3) and that there should be separation between policy and delivery. The policy made the distinction between urban areas, which need a framework for achieving efficient development and management of water supply and services, and rural areas, where communities were to be given responsibility for water services. The aim was for water users in rural areas to become the planners, owners and managers of their water services, making a contribution to the initial investment costs, and paying for operational and maintenance costs. In its 2003 Economic Survey [28], the Government of Tanzania reported that, by the end of that year, a total of 8363 rural water committees had been set up and 7226 associated water funds had been established.

This was good progress with the prospect of these initiatives turning into sustainable water services for the rural communities in Tanzania, but a study [29] carried out in 2005 by the University of Bradford suggests that community participation could be to the disadvantage of the poor. The study carried out on the Uchira system in Tanzania indicated that the water project was controlled by the elite in the village, and that water charges were seen as necessary for sustainability, to the neglect of providing the poor with their basic water needs. Prior to 2000, the village of Uchira in the foothills of Kilimanjaro did not have a piped water supply, which was of concern to the village leaders. With finance and advice from the German Government's GTZ, and using local labour, a 5.8 km pipeline was constructed to bring spring water by gravity from the Lyambala catchment to the village; 181 private houses and 13 commercial premises were connected through a distribution system. All supplies are metered and GTZ reports 90% revenue collection efficiency with some consumers either unable or unwilling to pay their water bills [30]. For non-payment there is a policy of disconnection and follow-up. The scheme is managed by the Uchira Water Users Association. Depending upon how the Users' Association handles the non-payments, there could be concern for the poor, but a greater concern was the lack of attention to the quality of the drinking water, and to basic hygiene. A study [31] carried out by a student from Yale University reported that, although the introduction of the piped supply had resulted in some reduction in illness, health workers were still dealing with a high incidence of dysentery. GTZ, on receiving this report, made provision for the installation of chlorination disinfection and funded a progressive health education programme. Also, The Government of Tanzania recognized, in its 2005 "New Water Strategy", that it needs to give greater attention to sanitary services. But can the Uchira project be considered a success or not? Clearly, the villagers are a lot better off than they were before, when women had to collect water from a distance, and the piped supply is making much more water available. The introduction of chlorine will provide for a much safer water supply, as microbiological quality of water is paramount (see Chapter 4). The GTZ report shows that the Users' Association has established a stable financial situation, without which there cannot be a sustainable water service. The Bradford University study points to some important factors in the organisation of community groups, but we should remember the advice of Voltaire (1694–1778), a strong defender of civil liberties, when he said "the best is the enemy of the good".

Above, Uchira is given as one example in Tanzania but in 2007 it was reported [32] that 8394 of a total of 10,639 villages had formed water committees to handle water and sanitation. These committees manage water and sanitation and are expected to meet all the costs of operating and maintaining their water supply systems through charges levied on water consumers. The government provides block grants to finance infrastructure. Tanzania is providing a large scale test-bed on community managed water and sanitation.

EFFECTIVE PUBLIC PARTICIPATION

The examples given above show the value and importance of public participation in the provision of water services, without which there cannot be understanding and support for necessary policies and actions. The examples also show that the level and type of participation is dependent on whether or not there is already a good water service, and dependent also on the size of the community and the complexity of the water service operations. Crucially, it is dependent upon whether or not there is already a degree of transparency. Therefore, I draw my conclusions first on public participation in developed countries, and then in the developing countries making distinctions between urban and rural communities.

Developed countries

Generally in most of Western Europe, in much of North America, and in countries like Japan, water services are well-developed, and the general public are not clamouring for direct control of them. Most people do not take much interest until they receive a bill or something goes wrong. They expect to be able to turn on the tap and receive clean water, and flush the toilet without thinking about how that water service is managed. Equally, the media take little interest when everything is working well. The situation changes when there are drought conditions, and water usage is restricted, or when there is a pollution incident. But by and large, "out of sight and out of mind" applies most of the time. On the other hand, the public expect to receive accurate and timely information to give reassurance on drinking water quality and on protection of the environment. The public become very concerned if they suspect that something is being hidden, or if they feel that they are paying excess charges for their service. The latter becomes an issue in situations where the water service provider is from the private sector. They do not like to see companies making profits at a time when they have a restricted water service, not recognising that profits are necessary to pay for investment, and are the driver for greater efficiency. This important issue is discussed in Chapters 8 and 9, but is mentioned here because of its relevance to public attitudes and understanding. The public appear to believe that good reliable independent information and transparency are even more important when private companies are involved. I believe that there is an equal requirement for transparency with public sector operations, as with less separation of policy and delivery, there are greater opportunities for hiding inefficient operations. So whether with public or private delivery, I see Consumer Councils as a necessary part of achieving transparency, as most members of the general public have neither the time nor the knowledge to be able to scrutinise published data. There is a need for a dedicated independent professional body to act on the public's behalf, and I see the UK Consumer Council for Water as a highly effective model.

Developing countries – rural areas

In rural areas, with relatively small communities compared with urban situations, as shown by the examples given in this chapter, it is possible for people to be involved directly in decisions on the provision of water services, and for communities to be organised so that they can provide much of the labour in the construction and operation of those services. There are many successes, often through the facilitation of NGOs who provide guidance and who assist in obtaining financial support. History suggests that direct replication of effective examples may not work, as success is highly dependent on local conditions. However, based on examples with positive results, I see the following ingredients for success:

(1) Government support for local initiatives, as in the case of Tanzania, although some examples in India show that, with NGO help, it is possible to make some initial progress without government support and, as a result, stimulate government to take an interest. Part of government's policy role is to provide the necessary legal framework to give communities security of tenure on their systems.
(2) Strong and effective community leadership to recognise the need for action, and with the vision to see what assistance is required, as in the case of the Alwar district in India.
(3) A respected NGO with the knowledge to facilitate projects, provide training, and give policy, technical and financial guidance.
(4) Choice of "technology" which allows for local production and maintenance and which provides not only the required quantity but a microbiologically safe drinking water.
(5) A policy of charging for water, leading to full-cost recovery for sustainability, but with provision for the poor as in Kolhapur in India. Full-cost recovery, with financial provision for refurbishment of systems is essential for long-term sustainability, and it also engenders a sense of ownership and responsibility. It is essential that attention is not given to those well-meaning people who promote the view that the poor cannot afford to pay for water. Those poor, without an official supply, have to afford much higher charges through purchase from water vendors. What the poor cannot do is save to pay monthly bills, so provision for them to "pay as they go" is needed. This is discussed fully in Chapter 3.

Developing countries – urban areas

There are two difficulties in urban areas. One is the large size of communities making it much more difficult to establish coherent community groups, a problem that is compounded in peri-urban areas in which communities have grown rapidly and haphazardly, without the benefit of established community spirit and values. The other difficulty, in terms of water supply, is the practical one of the lack of availability of suitable water sources and insufficient land for large-scale rainwater harvesting. This places much greater reliance on outside government or other institutional involvement in the provision of the water. The Savelugu case in Ghana has worked well but it was a long-established community. Even there it has needed outside help in the provision of a reliable water source.

The Orangi Pilot Project has shown what can be achieved by operating with small street units recognised by local communities. Clearly this requires much NGO effort and a lot of time. The lessons learnt include the need for long-term and consistent advice, patience and transparency. Success at Orangi has provided the impetus for other communities to develop and operate systems with over 2 million people now served by that approach. It is good to learn that OPP-RTI knowledge and experience has been used nationally through involvement in formulating the National Sanitation Programme in Pakistan. Referring back to Sanitation 21, described in Chapter 1, there is a need for governments to recognise the multi-layered nature of the challenge, with the requirements of each layer needing to be satisfied. This requires urban development planning to include government-led provision of the "big" parts of the infrastructure which take care of the environment, but coupled with support for communities in the development and execution of their local policies and systems. This necessitates effective training and capacity building at government level, with good NGO support on technical and social aspects at the local community level. Approaches to implement sanitation as a human right are discussed in Chapter 1b.

THE ROLE OF GOVERNMENTS

Effective public participation in developing countries depends critically on government policy in establishing a framework in which there are sound overall policies, as in Tanzania, to stimulate the involvement of communities, and to provide both financial and capacity-building support. What can people do to ensure that such policies result in their obtaining the necessary water and sanitation services?

WaterAid, an established UK NGO with extensive experience in developing countries, believes that an important aspect has been missing in the development of water and sanitation, and this is government accountability to the people. WaterAid coordinates a major initiative involving local NGOs called Citizen's Action which, to quote from its March 2006 report [33] on progress, "helps communities engage with service providers and government and then supports that engagement for as long as required". WaterAid stresses that the local organisations facilitate the process rather than mediate on behalf of citizens. The approach is to support citizens in dialogues and negotiations with service providers and governments. The aim is to "bridge the gap" between promises and reality. People have become involved in large numbers, and generally water service providers see the process as helpful rather than as a threat. Governments, in developing their national water plans, as emphasised in the UN Human Development Report 2006 [34], can benefit by supporting and encouraging this process.

The 2008 WaterAid report [35] reviewing progress on the citizen action approach shows how it has gone back to the basics of citizens needing to understand what is the water service like now, what should it be and who is responsible for providing it. With this information citizens are in a position to inform service providers of the gaps in service delivery and to help by suggesting solutions. Community score cards which provide the means of obtaining that information are able to identify different views within a community. Experience with the citizen action approach has highlighted that although citizens' involvement is essential, alone it is not sufficient. Often local governments do not have the resources or the skills to be able to respond effectively. Success requires linking citizens' action with the development of local government resources and capability. This is one of the key messages in Chapter 1b on the implementation of water and sanitation as a human right.

THE BEST IS YET TO COME?

When Shelley Arnstein developed her public participation ladder in 1969, it was unlikely that she could have envisaged the development of computing and the worldwide web. Database systems accessed through web communications provide the opportunity for public participation on a very local scale, and also on an international scale. World Water Monitoring Day (WWMD), now organised jointly by the International Water Association (IWA), and the Water Environment Federation of the United States (WEF), involves river and lake quality monitoring by community groups in over 50 countries. Around the date of 18 October every year, community groups use sampling and analysis kits to measure water temperature, dissolved oxygen, pH and turbidity, which give a spot indication of the condition of the water environment. The data are sent to the database on the WWMD; see IWA and WEF websites for details [36]. Although not a rigorous scientific approach, people are able to obtain an appreciation of how their water environment compares with elsewhere in their country, and elsewhere in the world. This is a transparent system, which attracts media attention, and provides the means by which the public can put pressure on governments to control pollution and make improvements.

Stephen Johnson [37] supports systems at the local level, like "311" in New York, which is a community phone-in number for anything to do with local government services. This could be developed to provide geographical information systems (GIS) accessed by both "experts" and amateurs. Johnson had in mind how such a system could have assisted John Snow and Henry Whitehead in speeding up the process of identifying the source of the London cholera outbreak of 1854. Such studies require specific local information, which apart from statutory medical returns, can be provided only by local people. Such systems could be powerful in other ways, such as collecting information on points of likely water distribution leaks, and also public monitoring of increased access to services and illegal connections. This raises the question of whether such systems are likely to be available in the developing world. Given the speed at which mobile phones have spread to all parts of the world, and the convergence between phones and computer technologies, it could happen very quickly. Local NGOs are in a position to promote this form of public participation, and to provide the "educated amateur" expertise.

SOCIAL NETWORKS

When in the first edition of this book I said the "best has yet to come" I didn't envisage that social networks would have become so influential. Woe betides any political party which doesn't recognise the important role of social networks in democratic elections. Social networks have become a means whereby oppressive States have greater difficulty in suppressing information. They provide a rapid means of transmitting information within professional and other groups. They can be very helpful in providing information during "disasters" as citizens can share useful information, but also with the risk of also "broadcasting" misinformation – an example of this was during Hurricane Sandy in the United States [38].

There are many examples of the use of social media. Here are some. WaterAid [39] and water-based consultancies [40] link up with Twitter, Facebook, LinkedIn and other social media. Such media are used to give awareness of events such as World Water Day [41] and of activities linked to it. Thames Water placed a video: "Sewerman Style 2012" on YouTube. The video [42] with Thames Water workers doing their own version of "gangnam style" in London sewers graphically publicises the problems of blocked sewers arising from cooking fat. There is an example of a documentary film company [43] promoting a film (Global Water Crisis) through mobile phone networks. The plan was to use a mobile phone app to calculate how much water is used in various actions in daily life. Depending on the accuracy of the information and the motives, the app example is a force for good or misinformation. Mobile phone-based water bill payments (See chapter 3) also provide the prospect of greater transparency and less corruption. There are examples of blogs [44, 45] being used to put pressure on local governments to deal with pollution. The United States Environment Protection Agency (EPA) includes the use of social media in its Community Outreach Toolkit [46]. The influence of social networks can only increase. This will require official information to be timely, accurate and available through social networks.

REFERENCES

1. Hochstetler K. (2012). Civil society and the regulatory state of the South: a commentary. *Regulation & Governance*, **6**, 362–370.
2. Arnstein S. R. (1969). A ladder of citizen participation. *Journal of the American Institute of Planners (JAIP)*, **35**(4), 216–224.
3. Humphreys P. C. (2002). Effective Consultation with the External Customer. Institute of Public Administration, Dublin.
4. Gaventa J. and Goetz A. M. (2001). Bringing Citizens Voice and Client Focus into Service Delivery. IDS Working Papers 138.

5. Engaging Citizens in Policy Making OECD Policy Brief No 10, 2001.
6. Choosing a consultation method: http://www.local.gov.uk/ (accessed July 2013)
7. World Development Report (2004). Making Services Work for Poor People. World Bank, Washington DC.
8. Consumers Count (2008). How Water and Sanitation Utilities Can Become more Accountable to Their Users. Water Practitioner Note (PNote) Issue 15 World Bank, Washington DC, June 2008.
9. Ackerman J. M. (2005). Social Accountability in the Public Sector: A Conceptual Discussion. Report jointly commissioned by the Social Development Department, the Poverty Reduction and Economic Development Network and the World Bank Institute. Social Development Papers, Participation and Civil Engagement Paper No 82, March 2005.
10. Ravindra A. (2004). An Assessment of the Impact of Bangalore Citizen Report Cardson the Performance of Public Agencies, ECD Working Paper Series – No, World Bank. 12 June 2004.
11. Promoting Voice and Accountability in Urban Water and Sanitation Services (2011). Lessons on Introducing Demand Side Accountability Tools in Kenya. Water and Sanitation Field Note, World Bank, Washington DC, June 2011.
12. Subgroup Report on the Lowermoor Water Pollution Incident (2005). Committee on Toxicity of Chemicals in Food, Consumer Products and the Environment. Department of Health. Crown Copyright, UK.
13. Holding Public Meetings. Improvement and Development Agency for Local Government, website: http://www.local.gov.uk/ (accessed July 2013)
14. News report (2003). 18 February 2003, Ghana Homepage.
15. Apoya Patrick (2003). Paper prepared for the Civil Society Consultation on the 2003 Commonwealth Finance Ministers Meeting. Commonwealth Foundation.
16. Ghana Celebrates Guinea Worm Eradication: Article in Daily Graphic Ghana. 28 July 2011.
17. MiDA Meet The Press – CEO's Speech: Press Release from the Ghana Millenium Development Authority. 7 February 2012.
18. Community-led Water Management in the Alwar District, Rajasthan, India. United Nations Unversity Press.
19. Arvari Sansad (Parliament) The Voice of the Common People. Tarun Gharat Sangh, website. http://tarunbharatsangh.in/ (accessed July 2013)
20. Johad – Watershed in Alwar District Rajastran. Tarun Gharat Sangh, website. http://tarunbharatsangh.in/ (accessed July 2013)
21. Jal Bhagirathi Foundation website. www.jalbhagirathi.org (accessed July 2013)
22. Sustainable Community Management of a Multi-Village Water Supply Scheme in Kolhapur, Maharashtra, India Water and Sanitation Program Field Note SPI Series 2 DFID, 2000.
23. Hasan A. (2001). A model for government-community partnerships in building sewage systems for urban areas. The experiences of the Orangi Pilot Project – Research and Training Institute (OPP-RTI), Karachi. *Water Science and Technology*, **45**(8), 199–216.
24. Situation Report July–September 2012: Orangi Pilot Project Research and Training Institute (OPP-RTI), www.oppinstitutions.org (accessed July 2013)

25. Institutionalized Community-Managed Approach for Water Supply and Sanitation in Agusan del Sur (2006). Country Water Action. Asian Development Bank, Philippines, May 2006.
26. Institutionalized Community-based Approach for Water Supply and Sanitation in Agusan del Sur (2006). Country Water Actions. Asian Development Bank, Phillippines.
27. Ally, M. (2002). Challenges towards Community Management of Water Supply Systems in Tanzania. www.TANZANIA-NETWORK.DE (accessed July 2013)
28. Developments in Different Sectors of the Economy – Water (2003). The Economic Survey. Government of Tanzania.
29. Toner A. and Msuya E. (2005). The Illusion of Community Ownership: Community-Based Water Management in Uchira, Kilimanjaro Region, Paper presented to East African River Basin Conference, USA, Morogoro, Tanzania, pp. 7–9, March 2005.
30. GTZ Databank on Rural Water Supply.
31. Schewel L. (2005). You can't see them with your eyes: water quantity and water quality. *The Yale Journal of Public Health.*
32. Water Sector Performance Report for the Year 2007/8: Ministry of Water and Irrigation, United Republic of Tanzania. Sept 2008.
33. Bridging the Gap (2006). Citizens Action for Accountability in Water and Sanitation. WaterAid, London, March 2006.
34. Governance for Sustainable Human Development (1997). A UNDP Policy Document. United Nations Development Programme, New York, January 1997.
35. Stepping into Action (2008). The Second Report on Citizens'Action for Accountability in Water and Sanitation. WaterAid, July 2008.
36. Websites: www.iwahq.org.uk and www.worldwatermonitoringday.org (accessed July 2013)
37. Johnston S. (2006). The Ghost Map. Riverhead Books. Published by the Penguin Group, ISBN: 9781594482694.
38. Social Media's Role in Crisis and Misinformation. SocialMediaToday. 1 November 2012. http://socialmediatoday.com/dneisser/965591/social-media-s-role-crisis-and-misinformation
39. Find us on these social networks: http://www.wateraid.org/uk (accessed July 2013)
40. Community Water Preservation and Policy. CH2MHill Water Blog. www.ch2m.com/corporate/sr/society/community-water-preservation-and-policy.asp (accessed July 2013)
41. Video Thames Water workers do their own version of Gangnam Style in London sewers. Article in London Evening Standard.
42. World Water Day: How Social Media Campaigns are Making a Splash. http://mashable.com/2012/03/22/world-water-day-2012 (accessed July 2013)
43. Documentary and Mobile Campaign Tackle Global Water Crisis. http://mashable.com/2012/03/14/last-call-at-the-oasis (accessed July 2013)
44. Come on in, the water is toxic: USA TODAY, February 21, 2013.
45. Citizen Pollution Reporting Leads To Cleaner Water: Blue Water Baltimore Water Blog. 9 January 2013.
46. Community Outreach Toolkit: USEPA EPA-840-B-12-001, Washington DC, http://water.epa.gov/polwaste/nps/outreach/upload/EPA-Community-Outreach-Toolkit-508-Compliant.pdf (accessed July 2013)

Chapter 6

Stimulating improved performance of water service utilities

How to encourage improvement

Conditions have to be appropriate for the establishment of effective and efficient water services. It is difficult, and probably impossible, for effective and efficient services to exist within a traditional local authority structure, in which there is often a lack of clear management accountability, a lack of separate and transparent accounting, and where the officials are responsible for both policy and service delivery. The importance of separation of policy and service delivery functions has been covered in Chapter 1, but this is so important for successful water services that the main points are repeated below. Given separation, it is also necessary to provide incentives for service delivery managers to perform well. These are covered below under the headings of "Management incentives" and "Benchmarking and comparative competition".

SEPARATION OF POLICY AND SERVICE DELIVERY FUNCTIONS

Government's role is policy making, and those involved in Government are generally most suited to a policy function. Elected officials are interested in politics and policy and, generally, local authority staff have been trained as administrators, rather than operational managers. There are exceptions, but generally governments are not good at managing operations, and with policy responsibilities having to take priority, the direction and energy of the elected officials is inevitably taken up in local political matters. The loser is efficiency of utility operations. The solution is to separate the delivery functions, so that good operational managers, free from local politics, can be employed to focus on efficient delivery of services. Essentially, there are three ways of achieving separation, all of which have been shown to be successful. These are internal contracts within municipalities, the formation of public-owned companies, and contracting out to the private sector. The key ingredients of each form of separation are transparency of budget and accounts, published targets and effective monitoring and reporting of the required deliverables.

BUILDING MANAGEMENT CAPABILITY

There are examples of successful reform driven by good management. Two examples of this, Phnom Penh and NWSC Uganda are given in Chapter 7. So surely, some people might say, it is simply sufficient to improve the management of water service providers. However the process of improving management performance is very dependent on other factors including the starting conditions. In particular, it is dependent on the extent of political interference, and thus lack of operational autonomy, and on the limitations imposed by the financial and cost structures. In Chapter 7, I give another example, that of Haiphong Water Service Company in Vietnam. This involved a 15-year input by a donor agency, Finnida, to improve the performance of the company. The work started from a low base of competence, due to the existence of "vicious circles", the downward spirals resulting from low cost recovery, subsidies and poor pay (Figure 7.5). The project placed emphasis on building management capability and, importantly, confidence, and was very successful. It is a model which should be followed where there are similar circumstances. However, it must be recognised that developing management within a poor institutional structure takes a long time. For more rapid progress, it is necessary to start with reform of the institutional structure, together with reform of the financial structure, to give existing or new management the autonomy and incentives to manage. Ideally, this should involve both separation of government from operations, and also some form of independent scrutiny, to ensure objectivity and consistency in enforcing policies and regulations.

MANAGEMENT INCENTIVES

In private sector operated water services, management incentives are inextricably bound to the incentives of their companies to perform. One example is the incentive-based water charging system in the UK in which the rpi-x formula gives encouragement for water companies to outperform. This is covered later in the next section.

In public sector operations, the market pressures against which managements are judged, and either rewarded or penalised (through loss of jobs), do not exist naturally, and have to be created in some other form. It is argued that in the public sector, managements are motivated by the desire and satisfaction of providing a good public service, and that private sector type rewards or penalties are neither necessary nor desirable. In some individuals this may be true, but having some form of linkage between performance and pay provides an overt measure and record of performance. To achieve this in practice requires transparency in the form of public reporting of performance against the required deliverables. The targets must be easily and accurately measurable, and for extra reward, should require performance above an acceptable norm. Establishing effective general management incentives requires a great deal of thought and care, which is beyond the scope of this book. It is for the reader to seek suitable publications, but I refer to three typical references ([1], [2] and [3]) at the end of this chapter, which give the principles and case studies of successful approaches. It is appropriate in this book to concentrate on performance measures that are peculiar to water service operations.

INCENTIVE-BASED WATER PRICING SYSTEMS

Incentive-based approaches are most appropriate in a private sector situation, and such a system is applied by the economic regulator in England and Wales. It can be applied in a corporatised public company situation, where profits are paid to the owning local authority, although there may not be the same pressures on management to outperform. Traditionally, even in the UK before 1989, an incentive-based system was not used, as the long standing statutory water companies (See Chapter 9 for the history) were regulated on a rate of return basis. So how does an incentive system work?

In concept, there is a simple formula, rpi-x, which is used to calculate the change in average maximum price a water company is allowed to charge. Rpi is the retail price index, to take care of inflation, and x is a required efficiency factor. If a water company achieves greater efficiency than x, the result is increased profit up to the end of the review period. At the end of the next review period that additional efficiency saving is transferred to the customer in lower prices, and the formula is applied again. The simple rpi-x applies only in a steady state operational situation, and adjustments have to be made for any changes, such as the cost of meeting higher standards, but the principle is the same, with a water company given an incentive to become more and more efficient. The approach places a big responsibility on the regulator to monitor water company performance closely, to check that companies are maintaining their infrastructures, and that they are not cutting corners operationally and putting consumers at risk.

In the normal market place, competition makes any form of price fixing unnecessary, but generally water services are seen as a monopoly. There are proposals in the UK [4] to introduce direct competition through common carriage, where companies compete for water service customers, but the water is supplied through a common distribution system. It is too early to judge whether it will be an attractive market place for both existing and new water companies, and whether it will result in lower prices and an equivalent or better service to customers. At this time, no other country is considering common carriage for water services, and it can be assumed that there have to be other means of simulating the market place. This leads on to benchmarking and comparative competition.

BENCHMARKING AND COMPARATIVE COMPETITION

Because of the use of the word benchmarking to compare the performance of organisations, which was not its original purpose, it is necessary first to define the context of benchmarking in this chapter. To distinguish between the two forms of benchmarking, the terms metric benchmarking and process benchmarking are now used.

Metric benchmarking. This is the process of comparing performance between organisations, in our case water utilities. Metric benchmarking provides comparative performance data and also shows trends in the performance of individual organisations over time. Metric benchmarking is achieved through the use of a small number of key performance indicators, sometimes compared with a model of the perfect enterprise. Metric benchmarking is undertaken at a high level ie at company or site level, and thus is not intended for consideration of detailed aspects of operations. The economic regulator in England and Wales, Ofwat, and VEWIN (the Association of Dutch Water Companies) use metric benchmarking to provide comparative performance measures between water utilities in those countries. Attempts are also made to make international comparisons.

Process benchmarking. This is more detailed than metric benchmarking and, as the name implies, measures the performance of each part of an organisation's operational processes. The aim, based on the result of quantified analysis, is to learn from best practice. Some systems go down to a very detailed level. For example, the system developed by the Water Services Association of Australia [5] includes 600 processes and subprocesses. Process benchmarking requires the management of a utility to decide on which aspects of its operations should be improved and then to look for the "best-in-class" on those particular aspects. Depending on the particular items for improvement, the best-in-class may be another utility, or for non-technical aspects, for example human resource policies and practice, it could be an entirely different organisation. The requirement is then to define the key performance measures, to find out by studying the other organisation what constitutes best-in-class practice, and to develop the utility's policies and procedures accordingly. A programme for improvement is established and progress monitored using the defined performance measures.

In practice, both metric and process benchmarking are required, as metric benchmarking identifies areas of poor performance where improvements need to be made, and process benchmarking is a management tool for achieving the necessary change. Figure 6.1 illustrates the roles of performance measures and metric and process benchmarking.

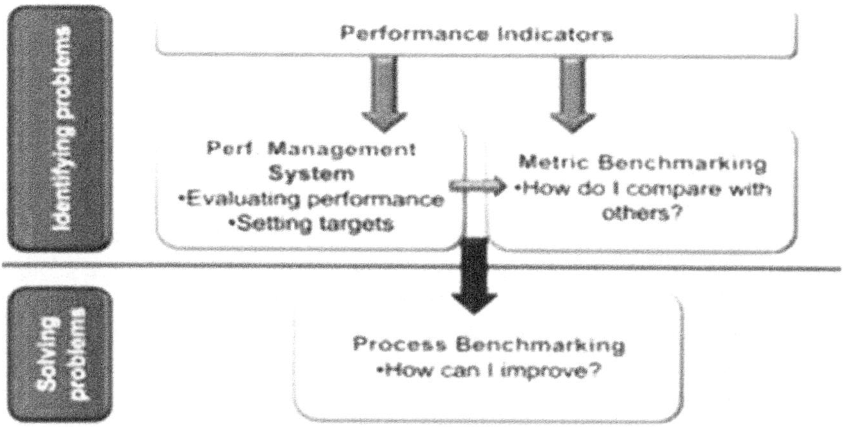

Figure 6.1 Roles of performance measurement, metric and process benchmarking (*Source*: [15]).

Metric benchmarking is a powerful tool for use by governments, by regulators, by utility associations and by a group of utilities, to stimulate change. However, it is necessary to have very clear outcome objectives and to choose performance measures that achieve those aims. Defining performance measures needs good knowledge of the operations. Governments and regulators, before finalising measures, should consider how they would respond if they were an operator. How would they, if they were operators, use the measures to show apparent good performance, and what might suffer as a result? A major difficulty is the inclusion of quality of service. A simple example is a utility's performance in response to telephone calls. It is easy to set a requirement to answer in a defined time or number of rings, but assessing the quality of response is much more difficult. It is necessary to avoid the situation where management can play the "numbers game". According to Pidd, [6] "performance measurement systems in public services can lead to dysfunctional consequences even when people operate with the best of intentions". In particular, Pidd's research into the use of public service performance measurement systems showed that if it is the motive of government, in using such systems to "control", rather than to stimulate, it leads to dysfunctionality. This is another demonstration of why there should be an independent regulator, free from political motives, whose role is to stimulate improved performance through carefully developed measures that produce the desired results. It is essential that performance management systems are developed in consultation with the operators, and piloted to assess any difficulties in measurement and data provision, before being rolled out across all operators.

To be effective it should be an open and transparent process with the results published in league tables, because rather like a football team, no utility likes to be bottom of the league. However, it is important that the published results are suitably qualified to take into account circumstances outside the control of a utility's management, for example source water quality. Alongside the league tables it is important to show trend information for each utility, because the improvement levels are more important than the position in the league table. To give an extreme example, a water utility at the top of the league table may have shown no improvements, or even slightly fallen back, whilst the one at the bottom may have made excellent progress from an inherited low performance position.

One difficulty in establishing a metric benchmarking scheme is the quality of available data, so it is important not to be too ambitious in the early stages. It is better in the beginning to have, for instance, just five reliable and accurate measures of relative performance, rather than twenty parameters based on poor information. However, the parameters chosen should be relevant to the key objectives. For example, if a key area is the level of revenue collection, then a simple measure is the percentage of billed amounts collected. Later on, if cash flow is important, then a time element covering time from meter reading to money in bank could be included. With objectives of improving revenue collection, improving standards of service and improving efficiency, a typical start up list of performance measures for a water supply utility might be:

- Money amount collected per 1000 metered consumers;
- Overall operating and maintenance costs expressed on the same basis to allow easy comparison of cost recovery;
- Pressure at the inlet to the meter;
- Overall water quality at tap;
- Staff numbers expressed as number per 1000 metered consumers;
- Leakage, initially as percentage of water going into supply, but with the longer term aim of using more useful comparison indicators, such as the Infrastructure Leakage Index (ILI).

Some of these items, for example drinking water quality and leakage control, are complicated, and require a good understanding of the technical issues. On leakage control the comprehensive book by Farley and Trow [7] includes the use of water balances to provide the critical understanding of where all the water is going.

Measures of drinking water quality compliance with standards can give an overall measure of drinking water quality based on compliance with the relevant standards; see Chapter 4 for the compliance figures for England and Wales, which show the impact of the regulatory enforcement system, and resultant improvement programmes. Another useful approach is to devise an operational quality index based on those parameters which reflect a combination of water companies' operational performance and consumers' perceptions of drinking water quality. The Drinking Water Inspectorate for England and Wales has used six parameters, faecal coliforms, turbidity, iron, manganese, trihalomethanes and aluminium, and has compared each individual company's performance against the average for all of the companies. This approach has the additional value of showing individual company relative performance on drinking water quality against a rising average.

EXAMPLES OF BENCHMARKING SYSTEMS

India

There were two pilot water supply and wastewater programmes involving the Water and Sanitation Program (WSP) World Bank. The first involved 13 water supply and sanitation (WSS) utilities and took place during 2003–4. The second pilot [8] involved 10 WSS utilities and took place during 2005–6. These pilots confirmed the benefits of benchmarking in identifying priorities for improvement and stimulating progress. Critically they identified the need for reliable information on all aspects of operational management. Poor data was weakening the ability to make objective decisions. Lack of sound data on production, coverage and metering were highlighted. Information on distribution networks was poor making it difficult to obtain reliable estimates of leakage. An example of the results from the second pilot is given in Table 6.1. Even without the guidance given on relative reliability of the data, the numbers point to questions rather than to meaningful performance information. This is not a criticism; on the contrary it is excellent that the participants were willing to take part and to expose their current situations as an important step in developing the service. I give this example because it demonstrates that introducing meaningful benchmarking is not easy. Records of installed infrastructure can be poor or non-existent and production and other meters may not be working and require repair or replacement.

The Government of India Ministry of Urban Development initiated an exercise to develop standardised service level benchmarks with respect to basic municipal services in the year 2006, covering water supply, sewerage, solid waste management and storm water drainage. A group made up from the Institute of Chartered Accountants of India (ICAI), the Public Record of Operations and Finance (PROOF), the Water and Sanitation Program (WSP) and Municipal Commissioners of Pune, Bangalore, Jaipur, Hyderabad and Kolkata was formed to develop a Handbook of Service Level Benchmarking. A draft was circulated widely among all the States and other key stakeholders for comment. A developed draft was produced and a National Level Workshop was held in July 2008 for the adoption of the benchmarks. The draft handbook was piloted in 28 cities, the data from which formed the Service Level Benchmarking Databook [9.] The Service Level Benchmarking [10] initiative was formally launched in

Table 6.1 Comparison of participating utilities on selected key indicators.

		Coverage %	Production litres per capita per day	Metering % of total connections	Working ratio opex as% of Revenue	Complaints % of water connections	Daily supply hours per day
City Boards	Bangalore	91	143	90	1.0	NA	2.50
City Boards	Chennai	98	107	4	1.4	44	3
City Boards	Hyderabad	95	192	93	1.1	39	1
City Company	Jamshedpur	79	608	1	0.9	43	6
City Corp.	Chandigarh	100	290	71	1.3	2	12
City Corp.	Indore	54	102	0	5.4	0	0.75
City Corp.	Pune	88	274	16	0.8	NA	7
City Corp.	Rajkot	98	126	0.4	6.6	30	0.33
State Agency	Bhubaneswar	45	269	1	3.3	8	3
State Agency	Dehradun	80	149	8	1.4	4	4
Reliability				A*	B	C	D*

Source: WSP [8].

* A refers to a high degree of reliability and D refers to low or negligible reliability. Darker the shade higher the reliability.

June 2010, and guidance [11]. based on experience with the draft handbook was published in September 2010.

The nine parameters chosen by the Government of India for water supply are listed below:

- Coverage of water supply connections.
- Per capita supply of water.
- Extent of metering of water connections.
- Extent of non-revenue water (NRW).
- Continuity of water supply.
- Quality of water supplied.
- Efficiency in redressal of customer complaints.
- Cost recovery in water supply services and
- Efficiency in collection of water supply related charges.

And the nine for sewerage are:

- Coverage of toilets.
- Coverage of sewage network services.
- Collection efficiency of sewage network.
- Adequacy of sewage treatment capacity.
- Quality of sewage treatment.
- Extent of reuse and recycling of sewage.
- Efficiency in redressal of customer complaints.
- Extent of cost recovery in sewage management.
- Efficiency in collection of sewage charges.

There are financial incentives for ULBs (Urban Local Bodies) to implement benchmarking. A sum of Rs 23,472 crores (~US $ 4.3 billion) is being provided by the Government of India for the improvement of basic urban services. The introduction of benchmarks is one of nine aspects of reform required for access to this fund.

England and Wales

An extensive benchmarking (comparative competition) system is operated by the economic regulator Ofwat with inputs from the quality regulators. Both the drinking water quality regulator, the Drinking Water Inspectorate and The Environment Agency issue separate reports on quality aspects. The regulation system is discussed in Chapter 9. Reports [12] produced by Ofwat on the high level performance scheme cover levels of service, overall performance, service related to consumer issues and delivery parameters for both the water service and the sewerage service. The results display data which has undergone quality control checks.

The parameters used reflect a different situation from India. As there is universal 24/7 coverage the emphasis is on consumer service and concern for the environment. The parameters used are listed below:

- Water supply (low water pressure, unplanned interruptions to supply, and drinking water quality);
- Security of supply (hosepipe restrictions, leakage, and performance against a security of supply index);
- Sewerage service (sewer flooding incidents and risk of sewer flooding);
- Consumer service (written complaints, billing contacts, billing metered consumers, telephone answering, telephone access, services to consumers with special needs, supply pipe repair policies, debt and revenue policies, complaint handling, compensation, and provision of information to consumers); and
- Environmental impact (sewage treatment works, pollution incidents from water and sewerage activities, and sludge disposal).

Ofwat has a scoring system allowing a comparative overall assessment across all of the parameters. The overall assessment for 2009–10 with a comparison with 2008–9 is given in Figure 6.2. Note that all companies score highly reflecting the impact of the regulatory regime and some achieve maximum (288) points.

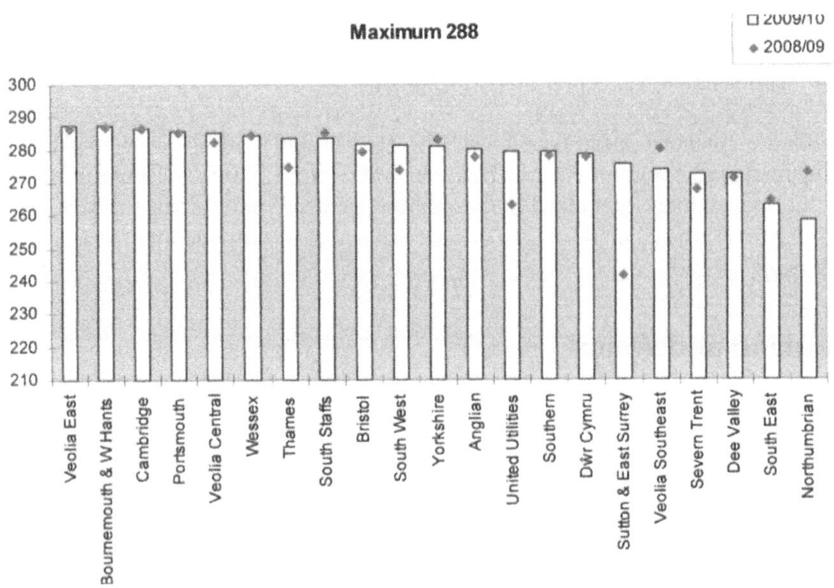

Figure 6.2 Overall performance assessment in England and Wales 2009–10 (*Source*: Ofwat [12]).

An important aspect of performance is care of the infrastructure in terms of serviceability. For underground water distribution systems leakage levels can give an indication of the state of systems. Ofwat uses a grading of improving, stable, marginal and deteriorating to show status for each company for both underground (infrastructure) and above-ground (strangely labelled as non-infrastructure), see Table 6.2. This is an important performance measure as it assists in determining the investment required in the infrastructure to maintain specified levels of service. The regulators have carried out extensive research to identify operational indicators which give early warning of deterioration. This is important as water infrastructure, particularly underground assets have long lives, and significant problems affecting service don't occur until there has been substantial deterioration. This is an area of regulation which is critical as it is unlikely that market forces alone, which tend to be shorter term focussed, would result in adequate investment now to prevent major re-investment in the future. The regulatory system in England and Wales over twenty years (1990–2010), of which comparative competition through benchmarking has performed an important part, has delivered greatly improved customer service, and has provided for investment in the "old" infrastructure following decades of neglect.

Table 6.2 Water service serviceability assessments 2009–10.

	Water infrastructure	**Water non-infrastructure**
Water and sewerage companies		
Anglian	Stable	Stable
Dŵr Cymru	Stable	Marginal
Northumbrian	Marginal	Stable
Severn Trent	Marginal	Stable
South West	Stable	Stable
Southern	Stable	Deteriorating
Thames	Stable	Stable
United Utilities	Stable	Stable
Wessex	Stable	Stable
Yorkshire	Marginal	Stable
WaSC assessment	**Stable**	**Stable**
Water only companies		
Bournemouth & W Hampshire	Stable	Stable
Bristol	Stable	Stable
Cambridge	Stable	Stable

(*Continued*)

Table 6.2 Water service serviceability assessments 2009–10 (*Continued*).

	Water infrastructure	**Water non-infrastructure**
Dee Valley	Stable	Stable
Portsmouth	Stable	Stable
South East	Stable	Stable
South Staffs	Stable	Stable
Sutton & East Surrey	Stable	Stable
Veolia Central	Marginal	Marginal
Veolia East	Stable	Stable
Veolia Southeast	Stable	Stable
WoC assessment	**Stable**	**Stable**
Industry assessment	**Stable**	**Stable**

Source: Ofwat [12].

Canada

The Canadian National Water and Wastewater Benchmarking Initiative arose from utility managers asking the questions "How well are we doing? How do we compare with similar organizations? Are we getting value for our money?" and "How can our utility begin the process of continuous improvement?" Around half, 43, of Canadian water and wastewater utilities, generally serving more than 50,000 people, take part covering around 60% of the population. The individual performance information is confidential to each utility but public reports [13] are made available. The benchmarking structure is defined by seven goals of a "Utility Management Model":

- Provide reliable and sustainable infrastructure.
- Provide accessible and sufficient infrastructure (capacity).
- Meet service and performance requirements at minimum sustainable cost.
- Protect public health and safety.
- Provide a safe and productive work environment.
- Have satisfied and informed customers.
- Protect the environment and minimize environmental impacts.

For each goal for each of water, wastewater and stormwater, there are a number of performance measures, which are extensive as can be seen in Table 6.3. Close attention is given to data quality with data collected by on-site visits from qualified and trained staff. It is required that any data input into the database is confirmed through a reference to the utility's financial records, annual report or other written records. There is an annual review of the management of the scheme.

Table 6.3 Canadian utility management model and number of performance measures.

	Utility management model – utility goals						
	Reliability and sustainability	**Infrastructure adequacy**	**Cost efficiency**	**Public health and safety**	**Safe productive work environment**	**Customer satisfaction**	**Environmental protection**
	Number of performance measures						
Water	12	4	23	9	8	2	3
Wastewater	18	3	37	5	4	2	3
Stormwater	4	4	3	4	3	2	3

Source: NWWBI [13].

The public report gives maximum, minimum and median for each performance measure. An example is given in Figure 6.3.

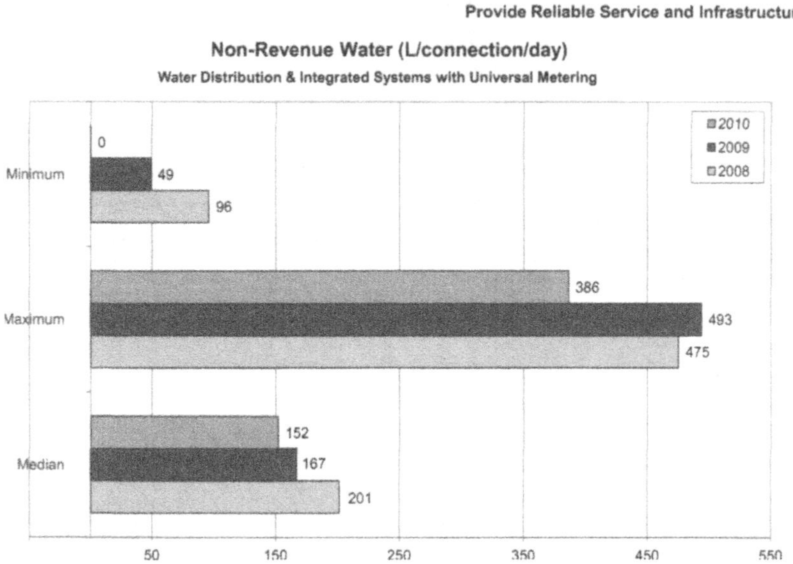

Figure 6.3 Minimum, maximum and median performance on non-revenue water (*Source* NWWBI [13]).

The Canadian system is robust with good data quality control and should provide utility managers with the means of mapping their own performance. The confidentiality approach has the advantage of making the scheme less threatening than one producing league tables, but it would be good if utility managers were able to identify the best in class for each aspect and learn from each other.

South Africa

In 2011 the South African Local Government Association (SALGA) and Water Research Commission (WRC) re-launched the National Benchmarking Initiative. A consortium of consultants which included a municipality were commissioned to build on the experience of an earlier initiative which took place over 2005–8 [14]. The initiative [15] is described as "for municipalities by municipalities to the benefit of municipal services" and seeks to introduce performance benchmarking in a bottom-up manner that (and I quote)

- Makes benchmarking part of "normal, good business practice" that assists officials with their day-to-day operations and demonstrates economic benefits and value to the water services sector.
- Focuses on hands-on support ("how do I do that?").

- Creates a support network and culture of information exchange between peers ("how did they do that?").
- Uses a web-based real-time data-capture and reporting system for tracking and measuring performance.
- Uses a modular, tiered approach to benchmarking to encourage and enable all to participate, at a level aligned with their current capabilities and future aspirations.

This is very much performance measurement and process benchmarking rather than league tables. There is recognition that there are large differences in the starting points across the various municipalities by the use of three tiers of complexity, basic, intermediate and advanced levels. Integrity of data is being given attention both for sound decision making but also to provide confidence in the system. A water manager was quoted as saying "Of course I want to see how I compare with others – for example, operating costs per kilolitre. But I wouldn't believe their data, because I doubt its integrity is any better than my own." It would be good to interview that manager after the scheme has been up and running for a few years to see if he then feels confident in the integrity of the data.

There is also a requirement for the municipalities to provide performance data to the Department of Water Affairs (DWA) for regulatory purposes. SALGA and DWA have agreed to coordinate the data collection formats to avoid duplication. This is an important step because it could be expected that the information required for regulatory purposes will be the same as that required by operational management to improve service.

National Water and Sewerage Corporation (NWSC), Uganda

This is not a country-wide scheme even though NWSC serves a large part of the country, but an approach used as part of reforming a previously poor performing large water and wastewater utility. A full account of the turnaround story is given in the book "Making Public Enterprises Work" [16]. A summary of the turnaround is given in Chapter 7. A key element in the approach taken was operational autonomy given to Area Managers with competition and reward stimuli for them to perform. Internal contracts defined the performance requirements. The Area Managers produced business plans for delivery of the required outputs. A department at NWSC headquarters, the Research, Monitoring and Evaluation Department, monitored performance [17] against requirements. Although the principle of non-interference in incentive-based performance management was adhered to, the monitors did not take a wholly passive role, as they would make flying visits to check data. Where they thought the Area staff were not aware of best practice elsewhere they would provide advice. This was important because one of the potential downsides of a competitive environment in this type of organisation is that there can be a tendency to avoid sharing better practice in order to retain competitive advantage. This negates against learning from best in class. In the NWSC case the Area Managers could choose whether to take such advice or try something else. They remained responsible for their Area's performance.

After the approach had been in operation for a while it was recognised that the performance requirements were largely related to quantity aspects, as these are readily measurable, and less attention had been given to quality performance. This is a common problem; at the time of writing attention is being given to the problems of quality of health care in England and Wales. To address this quality problem and to provide additional monitoring over and above the efforts of the small monitoring department, a "checkers system" was introduced by NWSC. Checkers from anywhere in the organisation would make unannounced visits to Areas, thus keeping all operational staff "on their toes". The trained checkers would cover all aspects of performance. The various departments produced checking manuals which would focus on aspects on which they had expertise. The engineering department checked on technical aspects, the finance department checked on billing accuracy and revenue collection and the human resource department checked on staff matters. All departments checked on qualitative aspects of area performance, namely appearance, cleanliness, maintenance, quality aspects of customer service, staff attitudes and behaviour, all of which impact on the image of the organisation. The introduction of checking required a balance to be made between monitoring outputs and inputs. After inspection checkers would announce their presence and conduct exit conferences. Findings and recommendations would be discussed and included in the checkers' reports. NWSC management considered the checking approach to be successful in accelerating the reform processes.

HOW TO START

The examples above, other than the Canadian initiative, involve largely metric benchmarking, with league tables used to stimulate performance through comparative competition. That is fine, but is it the best starting point? League tables can be threatening with the risk that organisations' efforts become devoted to the numbers game rather than improving the service. I see this in some government approaches which have been used in areas other than water services. As mentioned already, performance measures have to be chosen carefully, data quality is critical and should be subject to quality control, and critically the service providers have to share the objectives and approaches to improving performance. Buying in to the scheme in India was strengthened greatly through water service provider involvement in the pilots and in developing the handbook. I was involved in a pilot project in Egypt (not published) involving 7 water service providers to develop an approach to benchmarking. The pilot was based on non-revenue water and the first step was process benchmarking. This was seen as non-threatening and those involved were keen to learn from each other.

Interestingly each water service provider involved was able to contribute on some aspect on which they could be labelled "best in class". Those involved quickly took over the lead in the work, recognised the need for better data and made provision for improved data recording. By learning from each other they were able to assess their own relative performance and take steps to improve. The next step would be targeting year on year improvements. This is important because relative performance has to be qualified by different starting conditions and different operational circumstances. The key requirement is for each water service provider to make year on year improvements quite independently of others. Ultimately league tables can be used. However transparency is important and performance data should be available to the consumers. Even if a service is still relatively poor, year on year improvements are a powerful indication of intent and commitment to improved service.

However, an approach to be taken is highly situation dependent. In the Ugandan case, all the units involved were part of the same organisation in which competition was used as the main motivator for development. In that case the most threatening aspect might have been the introduction of checking but by that time the competitive culture had already been established. Management wouldn't have had any favourites and would have wanted all to succeed to the benefit of all of NWSC's customers. In cases where many different utilities are involved a slower transition may be more effective.

IBNET, the International Benchmarking Network for Water and Sanitation Utilities [18] set up by the World Bank, provides advice on setting up metric benchmarking systems and on promoting international comparisons. IBNET, and I quote from its literature, "provides direct access to the world's largest database for water and sanitation utilities" performance data, and supports and promotes good benchmarking practice among water and sanitation services by:

- Providing guidance on indicators, definitions and methods of data collection;
- Facilitating the establishment of national and benchmarking schemes;
- Undertaking peer group performance comparisons;
- Establishing links between utilities, utilities associations and regulators.

There are also excellent reference books, *Performance Indicators for Water Supply Services*, *Performance Indicators for Wastewater Services* and *Process Benchmarking in the Water Industry* ([19], [20] and [21]). Much of the literature is based on highly developed utility systems. For those less developed situations, often with poor information, it is important to start slowly with a few aspects and concentrate effort initially on obtaining good baseline data.

It is worthwhile repeating that making comparisons within a country is difficult, and results need to be qualified by giving the different circumstances of each company. Making comparisons between water utilities in different countries is even more difficult, and comparative data need to be considered carefully against the history, the backgrounds, and the conditions pertaining today. However, such comparisons allow governments and utility managers to ask questions about why another utility is apparently performing better in a specific area, to find out more, and to pursue classic benchmarking (ie process benchmarking) by learning from "the best-in-class". Metric benchmarking is a powerful tool because no water utility manager wants to be bottom of the league table, though to me the greatest value comes not from comparing actual performance, but relative improvements in performance. But is all this externally imposed pressure for performance really necessary? What about the old-fashioned public service ethic?

A PUBLIC SERVICE ETHIC – GOOD OR BAD?

Historically, there has been a tradition for those involved in water service provision to be motivated by the recognition that their jobs are providing for the good health of their communities, the so-called public service ethic. Undoubtedly this has been beneficial to public health, with water service leaders highly motivated to improve water quality and safety, and in developed countries, to maintain the supplies 24 hours of every day excepting burst water mains. These same people were also motivated to prepare for growing water demand and extreme events, such as droughts and floods. However it can be argued that the associated tenure for life, with managers remaining in the same jobs until retirement, resulted in inefficiencies due to an over-comfortable existence. Although there were cost pressures, there could be resistance to cutting costs through the sole argument of not putting public health at risk. Also, there was little transparency to the public, and an attitude by the engineer/managers of "we are experts and we know what is good for you" and, at that time, this was accepted by most consumers.

As the general public became more knowledgeable about health issues and the environment, there was an increasing demand for information and a demand to be treated as customers, rather than consumers. Governments responded, for example in the UK, through the use of Customer Charters. In the UK, at the launch of the Citizen's Charter in 1991, the Prime Minister at the time, John Major said, "The charter programme will find better ways of converting money into better services". Those public bodies meeting the charter criteria were awarded "charter marks" which they could display in their literature. As a result, the water service utilities had to respond in business terms, and their efficiency came under scrutiny. Transparency demanded published standards and compliance with them. Reliance on a wholly public service ethic driven approach was replaced by defined standards and regulation. This was necessary, and has achieved much higher quality standards much more efficiently, but often those achievements have been made by those same people who grew up with the public service ethic tradition. So for a period, there has been the best of both worlds, public-minded people working within a regulated business-orientated environment. The question for the future, with the emphasis on running water utilities as efficient businesses, is whether future generations of water utility leaders will see their jobs as protecting public health. If not, public health protection will become totally reliant on the quality of regulation. I would prefer to see a situation in which the water utilities take the primary responsibility for water-related public health, with good quality management and staff operating within a culture based on a public service ethic. It is good news, in this respect, that the emphasis on protecting drinking water quality has moved from end-of-pipe testing to managing risk (see Chapter 4). However, because consumers now see themselves as customers, a public service ethic is no longer enough. Customers of a monopoly service wish to have independent scrutiny of the performance of the service. Regulators provide that independent scrutiny, but their role should not be limited to checking that the water utilities deliver the required quality of water service. They should also encourage a public health ethic. Legislation can be helpful here if it includes penalties for a general lack of attention to safe operations. I like particularly the provision in Canada of "failure to achieve a standard of care" [12]. So in summing up - yes a public service ethic is good and necessary, but it needs to be reinforced by governance of operators in a climate of regulated business efficiency.

REFERENCES

1. Merchant K. and Van der Stede W. (2007). Management control systems performance, measurement, evaluation and incentives, 2nd edn. Prentice Hall, 2007. ISBN: 9780273708018.
2. Swanson R. (2007). Analysis for Improving Performance (English) (Tools for Diagnosing Organizations & Documenting Workplace Expertise), 2nd edn. Berrett-Koehler Publishers Inc, San Francisco. ISBN: 9781576753415.
3. Hale J. (2006). The Performance Consultant's Fieldbook (Tools And Techniques for Improving Organizations And People). John Wiley & Sons Inc, ISBN: 9780787985349.
4. Draft Water Bill. 10 July 2012 The Stationery Office, ISBN: 9780101837521.
5. National Performance Framework. Water Services Association, Australia, http://nwc.gov.au__data/assets/pdf_file/0020/11747/0607-National-Performance-Report-Indicators WSAA-PUB.pdf (accessed July 2013)
6. Pidd D. (2005). Perversity in public service performance measurement. *International Journal of Productivity and Performance Management*, **54**(5/6), 482–493.
7. Farley M. and Trow S. (2001). Losses in Water Distribution Networks: A Practitioner's Guide to Assessment, Monitoring and Control. IWA Publishing. ISBN: 9781900222112.
8. Benchmarking Urban Water Utilities in India II: Report, WSP World Bank, Delhi, September 2008.
9. Improving Service Outcomes 2008–9: Service Level Benchmarking Databook. Ministry of Urban Development, Government of India. September 2010.
10. Improving Urban Services through Service Level Benchmarking: Ministry of Urban Development (2010). Government of India, June 2010.

11. Handbook on Service Level Benchmarking: Ministry of Urban Development (2010). Government of India.
12. Service and Delivery, Performance of the Water Companies in England and Wales 2009–10. Crown Copyright 2010. ISBN: 9781904655787.
13. 2012 Public Report: National Water and Wastewater Benchmarking Initiative. http://www.nationalbenchmarking.ca/public/docs/Public%20Report%20_2012.pdf (accessed July 2013)
14. Review of the National Water Services Benchmarking Initiative (2010). Water Research Commission (2010) Project No: K8/926/3 (March 2010).
15. Moraka W., Bhagwan J., Mackintosh G. and Stevens F. (2011). Improving Water Services Provision through the National Municipal Benchmarking Initiative. Workshop at the 75th pre-IMESA Conference, Johannesburg, 2011.
16. Muhairwe W. T. (2009). Making Public Enterprises Work; From Despair to Promise: A Turn *around* Account. IWA Publishing.
17. Mugisha S., Berg S. V. and Skilling H. (2004). Practical Lessons for Performance Monitoring in Low-Income Countries: The Case of National Water and Sewerage Corporation. IWA Publishing, Uganda, Water 21, October 21 2004.
18. The International Benchmarking Network for Water and Sanitation Utilities (IBNET) www.ib-net.org (accessed July 2013)
19. Alegre H., Baptista J. M., Cabrera E., Jr., Cubillo F., Duarte P., Hirner W., Merkel W. and Parena R. (2006). Performance Indicators for Water Supply Services. 2nd edn, IWA Publishing, Jul 2006, ISBN: 9781843390510.
20. Matos R., Cardoso A., Ashley R., Duarte P., Molinari A. and Schulz A. (2003). Performance Indicators for Wastewater Services. IWA Publishing, Oct 2003, ISBN: 9781900222907.
21. Parena R., Smeets E. and Troquet I. (2002). Process Benchmarking in the Water Industry. IWA Publishing, Mar 2002, ISBN: 9781843390107.

Chapter 7

Steps for successful public sector operations

Reform is in order

The development of public sector operations requires far more than a "dose" of benchmarking, as fundamental changes are required in approach and structure. These fundamental changes can be made through initiatives within a city, as in the case of Seattle, which is discussed later in this Chapter, but generally they require action by national governments to stimulate reform, as in the UK prior to privatisation, and as in Australia and The Netherlands. To understand the fundamentals, it is necessary to consider the limitations of the traditional municipal model.

THE TRADITIONAL MUNICIPAL MODEL

The main argument for the traditional municipal model, in which water services are part of local government operations, is that it is close to the people, with responsibility resting with an elected mayor, and civic pride acts as a positive driver for good quality operations. The counter argument is that water services often lose out to other local investment, and there is little transparency with the result that water revenue may be used for other purposes. Clearly, there are good local authority operations, and local government has often been the means of developing water services for all in a community. However, conditions are stacked against most local authorities being able to develop sustainable water services because of the following:

- Lack of separation between policy and operations. It is unusual to find managers who are good at both policy and operational management. Policy development requires an appreciation of the bigger picture, whereas operational management requires close attention to detail.
- Planning is an internal budgeting process, and decisions on investment in water services are likely to be wholly political decisions, based on

competing demands on available money, and concerns not to raise water service charges ahead of elections, rather than on objective consideration of the needs of the service.

- Inadequate investment in maintaining and refurbishing the infrastructure, resulting in longer-term problems in service quality.
- Because mayors are considered to represent the people, there is no direct public participation. For the same reason, there is a lack of transparency with the risk of bad news being "swept under the carpet".
- Often, a city operation is too small to attract good management and afford required scientific support.
- There is no independent scrutiny or reporting which are necessary to provide openness and transparency.

GOVERNMENT ACTIONS FOR REFORM

There is a need for reforms to address these fundamental issues. Change is always difficult, especially if it involves elected officials and senior staff having to relinquish control of operations, and submit themselves to independent scrutiny. For this reason, national governments should introduce either legislation, or incentives, for local government reform of water services. In the UK, there were a series of reforms starting in 1945, which achieved the necessary scale of operations for viability, achieved separation of policy and operations, and brought about full-cost recovery from charges. Chapter 9 is devoted to the story of England and Wales.

In The Netherlands in 1975, the national Government first encouraged water supply utilities to consolidate, and then required those that did not respond to integrate their operations, resulting in a reduction of the number of utilities from 94 to just 13. I return to The Netherlands situation later. In Australia, the Commonwealth (National) Government took a different approach, as part of a programme of reform of all public services. The Government provided financial incentives for states to achieve defined goals, but without specifying how it should be done. This approach has been highly successful, and later in the chapter I describe how the city of Brisbane in the State of Queensland, and the State of Victoria approached their reform of water services.

Reforms in the developing world have been successful when there have been enlightened governments, able and willing to gain the benefit from donor funding and expertise. I give four examples in this chapter, those of Phnom Penh in Cambodia, the National Water and Sanitation Corporation of Uganda, the city of Haiphong in Vietnam, and urban water services in Ghana. One common lesson arising from these examples is that reform takes time.

Reform can also be stimulated by an event, such as the Walkerton water quality incident in Ontario, Canada, when the Provincial Government took a very different approach, with the overriding priority of delivery of safe drinking water. This provided for a variant of the municipal model with a strong provincial oversight, which I also discuss later in the chapter. Before continuing, I explain the four basic models for water service operations.

THE FOUR BASIC MODELS

The four structures are defined by the degree of separation from government, asset ownership and whether the operator is public or private. At present, the degree of regulation also depends on the structure, but I will argue the case for equivalent regulation in all four structures. The four structures are illustrated in Figure 7.1. All four models are subject to general corporate legislation, and drinking water quality and environmental regulation, with degrees of independence. The municipal model has been considered above, with a lack of separation from government and no formal regulation. The municipal model is still the most common, and in France where there is a tendency to refer only to the PPP (public private partnership) model, around half of the operations are carried out by municipalities. To achieve separation of operations from the government policy function, whilst remaining a wholly public operation, there is a second model of corporatised public companies.

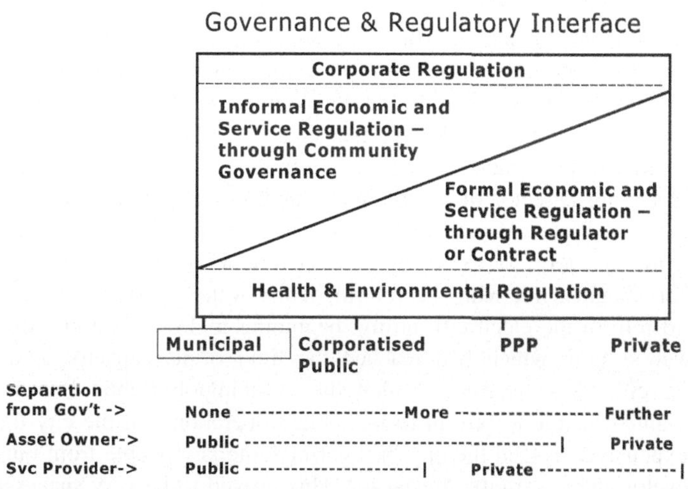

Figure 7.1 The four basic models of water service provider (*Source*: Paul Reiter).

Such companies are required to operate as if they are private companies, with private-sector type accounting and performance measures, but with public ownership, usually by local government. This is the model adopted by The Netherlands, the State of Victoria in Australia, and in cities in Germany and Scandinavia. Generally, there is no independent economic regulation, with tariffs being set by local government, although Victoria decided, after a few years into its reform programme, that an independent regulatory body should be established.

The third model is PPP, in which a municipality contracts out operations to a private company. This appears to have worked very well in France, although I have not been able to obtain performance data to compare with the other models. There has been mixed experience around the world, and I consider the reasons why, and the need for great care and skill in preparing PPP contracts, in Chapter 8. Generally, there is no independent regulation, with so-called "regulation by contract", the limitations of which I have discussed in Chapter 2.

Finally, there is the fourth model, in which the assets are owned by private companies, who both finance and operate the water services. Examples are England and Wales, which are covered in Chapter 9, and Chile which is described in Chapter 8, both having strong independent regulation. There are some examples in China in a less-regulated environment but, generally, there is an increasing amount of formal regulation in the move away from the municipal model to the other three models. In this chapter, I give examples of the reform of the municipal model, those of Seattle and Brisbane, and consider successes in the adoption of the corporatised-public model.

THE SEATTLE STORY

No, this is not about Boeing, nor Microsoft, but about Seattle Water. I should begin by thanking two people for their private communications, providing me with the background and information on the development of the water services in this city in Washington State, USA. My thanks go to Margaret Pageler, a Seattle Councillor and former Chair of the Water Committee, and to Paul Reiter, former Deputy Managing Director of Seattle Water and currently Executive Director of the International Water Association. I am privileged to tell the story in outline, but I hope that one day, someone will produce a book on the full story.

The story begins, not with the reform of the water operations, but with the decision to reform the electricity utility, beginning in 1985. The city operated a hydropower system, which had reached the limit of its capacity in a rapidly developing city. A consortium, called the Washington Public Power Supply System, wanted Seattle to buy in to its nuclear operation, but the city thought it was too expensive. Instead the city brought in some key people from outside the city to develop the electricity "business". This proved to be very successful, and the development team was moved to another part of the municipal operation, to consider the city's solid waste. The result of this work was the adoption by the city of a highly effective recycling system. So what has this to do with water? The connection is that during the reform of electricity and solid waste operations, this highly successful "internal management consultancy team", although also operational, had gained vital experience in turning round municipal operations, with all the inherent city politics. At that time in 1995, the city had been cited by the Environment Protection Agency (EPA) for failures in wastewater handling, and the "reform team" was moved to water operations.

In 1995, water and wastewater operations were separate entities, with wastewater coupled with drainage and roads. In a re-organisation in 1997, water, wastewater, drainage and solid waste were brought together under one management team. Prior to 1995, the city suffered from the common malaise of municipal operations, with councillors unwilling to increase water service charges, resulting in significant under cost recovery. There was severe under-investment in infrastructure, with the implied assumption that the distribution water pipes would last for 700 years. It was estimated that $1 billion investment was required to deal with the water infrastructure, including covering all treated water reservoirs and adding necessary water treatment capability. The new team put in place effective procurement systems, introduced longer-term capital planning, with an economic model to consider a set of alternative investment strategies and their impact on necessary cost recovery, and thus charges. The city embarked on a long-term water conservation programme, with a target of 1% reduction in per capita usage each year for ten years. The result was that Seattle was able to serve an additional 20% population within the same water supply level. In 2000, the city launched the Millennium "Salmon and Creek Clean-up Program". This involved significant additional investment but, through effective public participation, achieved strong public support. I said earlier that this story was not about Boeing and Microsoft, but the high investment programme in water and wastewater coincided with unprecedented wealth in the city, due to the success of those two large companies, together with many others during the "dotcom" boom. The consumers – and their elected Councillors - accepted that water and wastewater service charges would have to rise 10% each year in real terms for 10 years. So what was the structure that achieved this transformation, and how were the developments played out between the city and the management team?

The city formed a "Water Enterprise" within the city administration with accountancy laws ensuring that the water budget would be "ringfenced". The governance structure is as shown below in Figure 7.2.

Figure 7.2 Seattle City Water Services Governance Structure (*Source*: Margaret Pageler, Former Chair of Seattle City Water Committee).

Within that structure, the Mayor is the City Chief Operating Officer, the City Council is the Board of Directors and the Regulator, the Water Committee is responsible for policy and budgets, and Seattle Public Utilities is responsible for delivery of services. So there is separation between policy and delivery, although there is no form of independent economic regulation. However, in overcoming the inherent limitations of a municipal structure, Seattle recognised some key requirements for effective municipal governance, namely:

- Stable political institutions – this is dependent upon having elected officials, regardless of political party, whose sole interest in the water sector is providing effective, efficient and sustainable water and wastewater services.
- Availability of finance – the US tax incentives make municipal bonds an attractive route for municipalities. These bonds, being tax free, are attractive to investors, and provide municipalities with the ability to raise capital at relatively low fixed interest rates.
- Recognition by the community of the need for full cost recovery – this is dependent on effective communication with consumers.
- Transparency – the so-called "sunshine" laws requiring public decisions to be taken openly. Frequently, because municipalities are run by elected officials, and because water services are just one of the municipal functions, there is a tendency for communication with consumers to be seen as unnecessary.
- Means of preventing water income from being used for other purposes.

Critically, the service was designed for the benefit of consumers, with an operating contract as an organising principle. The contract with consumers is "to deliver clean water reliably at an affordable price". The contract with the community and the environment includes low-income subsidies, water conservation, forest preservation and salmon recovery.

So much for structure and intentions, but what has been achieved? The achievements are listed below:

- There is a high level of transparency through establishing good contact with the community and customers through public meetings, with a citizen committee on water charges, and an education programme on water conservation.
- A significant investment programme has been completed to meet drinking water standards and to guard against the threat from cryptosporidium.
- Seattle Water has achieved full cost recovery of capital and operating costs, including the investment necessary to refurbish the infrastructure.
- Asset management planning has been introduced to develop efficient and effective approaches on renewal of the infrastructure to achieve the required serviceability of the assets.
- With the need for water prices to rise by 10% each year for 10 years, to recover from decades of under-investment, due to under cost recovery, subsidies directed at low-income consumers have been introduced.

- There has been increased operational efficiency, without which water prices would have had to increase by even more than they have in practice.
- There is good collaboration with other agencies to achieve effective management of the whole catchment. Environmental objectives required by the community have been achieved. These objectives are beyond that required under federal law.

The question now for Seattle is whether the situation is sustainable. What is the risk of the city returning to its bad old ways of inadequate cost recovery resulting in inadequate investment, and a deteriorating infrastructure? During the reform period, the city had the benefit of a booming economy, driven by very successful industries. If the area were to experience a recession period, would local politicians yield to the possible pressures from consumers to cut water prices? It is useful to consider how well the Seattle model stands up to scrutiny against what might be considered to be the ideal governance model. It is large enough to afford good quality management and good scientific support. It has separation of policy and operations, but not regulation. It is self-financing through municipal bonds and has achieved full-cost recovery including that needed to refurbish and maintain the infrastructure. It is involved in a form of benchmarking through the international local government association, ICLEI. It has a form of time-limited licensing in that the MD of the Seattle Public Utilities has a period contract. However, it does not have an independent regulator who would objectively manage the rate setting process based on need, and there will always be the risk that politicians will have other motives or objectives. There was an instance where a local politician wished to use water revenue to finance a huge fountain in the harbour, but happily this was rejected. The architects of the city's reform are aware of these risks and have taken some steps to try to prevent inappropriate future policies. A set of codified economic and service standards were adopted by the City Council, and an Operational Board of the utility was formed with membership from surrounding utilities. Ultimately, whatever the safeguards, it will rely on the good sense and integrity of the elected officials.

A good indication of continuing support by elected officials in 2013 is agreement to ongoing tariff increases to pay for infrastructure development programmes. The approved tariff increases for the period 2012–14 are shown in Table 7.1.

Table 7.1 Approved Water Tariff Increases in Seattle.

Year	2012	2013	2014
Monthly Increase US $	2.25	2.43	2.55
Estimated Average Monthly Bill US $	37.25	39.68	42.23
Percentage Increase	6.4	6.5	6.4

Source: City Website.

BRISBANE, QUEENSLAND, AUSTRALIA

The approach taken in Brisbane in 1997 to reform was very similar to that of Seattle. There were major structural changes in water services in Queensland in 2011. I give a summary of both. I am grateful to Mark Pascoe, former Manager of Water and Sewerage of Brisbane City Council, and now Chief Executive of the International Water Centre, Australia, for his personal communications on the Brisbane history.

In 1995, Australia introduced the National Competition Policy, directed at the reform of the public sector, as part of developing the economy. It included the extension of "trade practice laws" to government businesses, and the reform of all laws that restricted competition unless, as in the case of water, it was considered that the benefits outweighed the costs. Price monitoring was extended to all government businesses with a market monopoly. The reforms, as applied to water, required universal water metering, full cost recovery and removal of cross-subsidies, the separation of water entitlements from land ownership, the provision of "environmental" water, the facilitation of water trading, rigorous appraisal of projects both economically and for ecological sustainability, improvements in water quality, separation of policy, a means of ensuring that utilities have a commercial focus for efficiency, and public education and consultation. In essence, the reforms addressed all the important fundamental elements for efficient and sustainable water services. States had the incentive of per-capita payments from the National Government for achieving the required reforms, with progress being overseen by the National Competition Council. The Commission produced annual reports [1], giving the progress of each State in achieving the required reforms.

The National Government did not specify how each state should approach reform, and a number of different approaches have been pursued. The city of Brisbane, in the State of Queensland, decided to pursue a market approach within a municipal structure; Figure 7.3 illustrates the approach.

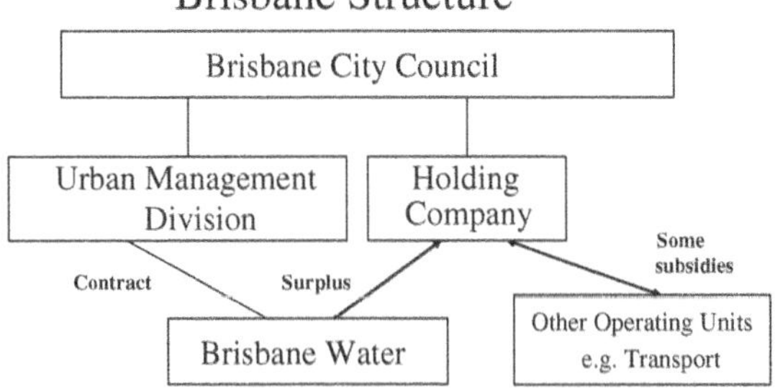

Figure 7.3 Brisbane City Water Governance Structure (*Source*: Mark Pascoe).

To avoid paying corporation taxes, Brisbane chose not to adopt the corporatised-public model, but to achieve a market approach by having internal contracts. Prior to the introduction of the national reform process, the city had already separated the Water and Sewerage Department from the main city functions. In 1997, a formal structure was adopted with the establishment of Brisbane Water, as a commercial business unit of Brisbane City Council. Initially, its Board was appointed by the City Council, led by a City Councillor, and included "directors" from the private sector. Later, the city established a Holding Company, made up of chairpersons of various business units, one of which was Brisbane Water, with additional private sector directors. A purchaser/provider model was chosen, with the purchaser being the Urban Management Division of the City Council. Within this structure, the Urban Management Division is the asset owner, and "purchases" the management of the operations through a management contract. The contract is intended to be long-term, but to achieve a licensing effect, based on management performance, the contract comes up for review every five years. The contract sets performance measures and the Water Department monitors the performance of the "contractor" against those measures. Investment is by the Urban Management Division with the assets leased to the contractor through a franchise fee, which is calculated on the basis of a return on assets. Initially, the required rate of return was 4% with the intention to increase it to 7%. The contractor, Brisbane Water, is wholly responsible for service delivery with no interference allowed by the Urban Management Division. Brisbane Water pays a "dividend" to the Holding Company, together with any other surpluses. The success of this approach is shown by significant improvements in operational performance, the details of which are published, in comparison with other Australian cities, by the Water Services Association of Australia [2]. A discussion paper [3] prepared for the Department of the Prime Minister of Australia, indicates that Brisbane has one of the highest levels of cost recovery among Australian cities, based on efficient replacement cost of the infrastructure, of 84%.

As with Seattle, it is worth testing the robustness of this structure with the "ideal" governance model. It brings a commercial approach with the benefit of Brisbane Water having a "business" Board, achieving separation of policy and delivery. However, there is no independent planning or rate fixing process to ensure total objectivity, free from political interference. With the Holding Company already using surpluses provided by Brisbane Water to subsidise transport and other service delivery in the city, there is the risk that the improved efficiency of operations will not benefit water consumers through lower charges, and that water could be used as a "cash cow". However, the approach yielded significant benefits to the city. As it has been shown to be an effective alternative to the corporatized-public model I retain the text on this valuable experience even though water services in South East Queensland changed after 2007.

As a result of the severe drought conditions affecting Australia, the State of Queensland rethought the provision of water services in South East Queensland, with restructuring legislation [4] introduced in 2007. There was separation of retail operations from bulk supply and Brisbane became one of five cities served by Queensland Water Utilities. The structure in 2010 is shown in Figure 7.4.

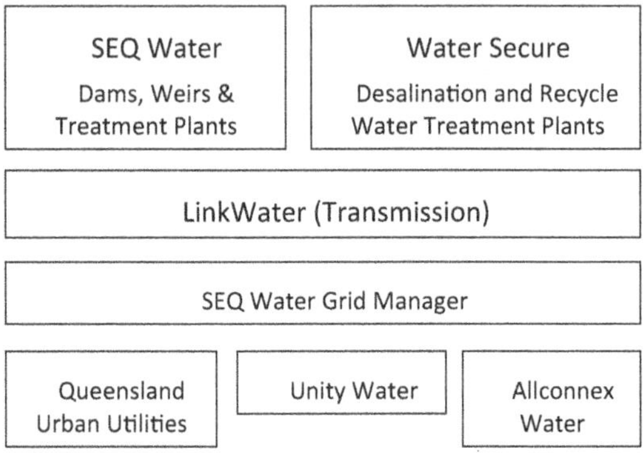

Figure 7.4 Structure of South East Queensland (SEQ) Water Supply 2010 (*Source*: State of Queensland [4]).

Late in 2010 the drought ended, followed by unusually high rainfall, such that the new (desalination and re-use) sources of water were not required. So in 2011 SEQ Water and WaterSecure were merged. In May 2012 there was a change of government in Queensland. Legislation [5] was introduced in 2012 taking out the trading aspects of the 2007 legislation, and adopting a structure (See Figure 7.5) similar to that in Melbourne, Victoria, which is discussed later in this chapter.

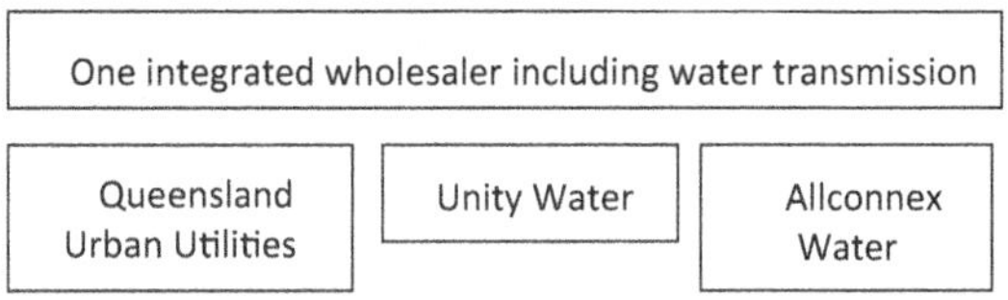

Figure 7.5 Structure of South East Queensland Supply 2013 (*Source*: State of Queensland [5]).

The economic regulator is the Queensland Competition Authority. Water quality and service regulation is carried out by the Office of the Water Supply Regulator, which is within a new state Department of Energy and Water Supply. A combination of highly variable rainfall and thus water security, coupled with different political views has resulted in substantial changes in approaches. This experience may be relevant to proposals for changes in the structure in England which are discussed in Chapter 9.

ONTARIO, CANADA

The Walkerton incident, which took place in the year 2000, was followed by a public enquiry under Justice Dennis O'Connor, resulting in recommendations for major changes in the governance of water supply services [6]. Walkerton is a small town with a population of around 5000 people, supplied by chlorinated well water. In May 2000, there was a failure of disinfection, and the water supplies contained the pathogens *E.coli 0157:H7 and Campylobacter jejuni*. About half the population of the town became ill and 7 people died. Justice O'Connor's report was critical of both the Walkerton Utility and the Provincial Government, referring to:

- Failures in basic monitoring;
- Lack of effective operating procedures;
- Lack of training of operators;
- Failures by provincial inspectors to notice deficiencies in treatment and monitoring;
- Failures in the provincial approvals and inspection programmes.

The main recommendations were:

- Public participation in setting quality guidelines;
- The accreditation of all water utilities;
- Provincial emergency response plans for all systems;
- Continuous monitoring of all supplies;
- Review of management and operational procedures by municipalities;
- Training of all system operators;
- The creation of a Provincial office of Chief Inspector, Drinking Water Systems.

The Government responded quickly to the report with the Safe Drinking Water Act 2002 [7]. The Act is very detailed, covering roles and responsibilities, licensing of utility owners, accreditation of operating authorities, licensing of laboratories, detailed requirements for operational plans, standards of care and penalties for lack of standards of care, and procedures for handling incidents. The Act was followed in 2003 by the drinking water regulations, which give detailed requirements on water treatment, sampling and analysis, and reporting. The Act places the responsibility for safe drinking water, not with the municipalities, but with the Minister for the Environment. The Minister is required to appoint

Directors to act on behalf of the Ministry, Provincial officials to investigate offences, and a Chief Inspector. The Chief Inspector's functions are to:

- Provide advice to the Minister;
- Implement operational policies on inspection;
- Periodically review regulations;
- Develop training programmes for Provincial Officers;
- Monitor inspection programmes;
- Produce an annual report covering the overall performance of the drinking water systems and the inspection programmes.

The Act sets down detailed requirements "for the preparation and content of operational plans, which are part of the process of an owner obtaining a licence. The licence can specify staffing levels and water treatment processes. There are strong powers to prosecute anyone who permits any "thing" to enter a system if this could result in a hazard to drinking water, contravention of a prescribed standard, interference with the normal operation of the system, or failure to achieve a "standard of care". Standard of care is described as exercising a level of care, diligence and skill. This is similar to the "due diligence" defence which water companies have in the UK, except that in the Canadian situation it is an offence.

If a Director is of the opinion that there is an imminent water health hazard, the Director may require one or more of following:

- Closure or restriction of operation of the system;
- Consumers be provided with an alternative supply;
- Consumers be notified;
- The system be operated in a way ordered by the Director;
- The water be sampled, tested and monitored in a particular way.

In effect, the Director takes the operational decisions. I am describing this Canadian system in considerable detail, as it demonstrates the extent to which a government can take central responsibility through establishing a strong top-down structure. The Chief Inspector's Report, published in May 2005, indicates that there have been improvements in drinking water quality compliance, and whereas 255 improvement orders were issued in 2003/4, the number in 2004/5 was down to 65. The Chief Inspector's Report covering 2004/5 [8], published in April 2006, shows further improvements with progress on resource protection through the implementation of what is called "a source to tap safety net". Progress over the period 2004–5 to 2009–10 is shown in Table 7.2. Clearly, the new regime has resulted in much safer water supplies, but the question is whether there may have been a better way of approaching the problems unearthed by the Justice O'Connor enquiry. One of the fundamental problems is the large number of small utilities working within a municipal environment. One approach would have been to integrate the small utilities into viable-sized corporatised-public companies, allowing the management to manage but with the Chief Inspector having a strong audit function backed by legislation.

Table 7.2 Performance in Ontario over the Period 2004–5 to 2009–10.

	2004–5	2005–6	2006–7	2007–8	2008–9	2009–10
Compliance %	99.72	99.84	99.83	99.85	99.87	99.88
Percentage of systems with operational performance greater than 90%	?	90	92	95	95	95
Percentage of systems with 100% rating	?	33	40	50	49	59
Improvement Orders	65 255 in 03–04	43	23	14	?	8

Source: Chief Inspectors Reports.

A report [9] by an Expert Panel, set up by the Ontario Ministry of Public Infrastructure Renewal, is critical of the current situation. In its 2005 report "Watertight – the case for change", the Panel refers to the problem of utility size, with 75% of systems serving less than 5,000 people. It considers water service charges to be very low, only 1/3 of that in European countries, producing insufficient revenue for maintenance of the infrastructure. The report describes the municipal structure as ineffective, with elected officials more interested in spending money on "more visible issues".

The Water Opportunities Act 2011 [10] does not address the issues raised by the Expert Panel, other than some encouragement for small utilities to work together. The Act sets the framework to help municipalities improve the efficiency of municipal infrastructure and services by:

- Identifying innovative, cost effective solutions for drinking water, sewage and stormwater system challenges.
- Optimizing systems and improving water conservation.
- Identifying opportunities to demonstrate and carry out new and emerging Ontario water technologies, services and practices.

As part of implementation of the Water Opportunities Act, a water technology hub, called the Water Technology Acceleration Project (WaterTAP) has been established. A description of WaterTap is given in its first annual report [11]: "WaterTAP was established to help grow globally competitive companies, and provide high-value jobs in Ontario's water and wastewater sector. It will identify opportunities, build partnerships, and improve information flow within the sector. WaterTAP will also provide advice to government on developing Ontario's water sector, with a goal of helping entrepreneurs access world-class research, talent and capital to create solutions for water management, treatment and protection." It will be interesting to follow this initiative to see if development of the utility supply sector will lead to changes in utility governance.

VICTORIA, AUSTRALIA

The background to the reform of public services in Australia has been described above in the section on Brisbane Water. Whereas Brisbane Water decided to reform within a traditional municipal structure, the State of Victoria took a different approach.

In 2004, Victoria had a population of 5 million people. Prior to the reforms, there were about 400 water utilities, many of which were very small. The State capital, Melbourne, was served by just one utility, but with privatisation in mind the operations were broken up into four corporatised-public companies, one wholesale and three retail utilities. Later, having achieved very significant improvements through the change to a corporatised-public structure, with private-sector business practices, the decision was taken not to involve the private sector. Throughout the rest of the State, there were widespread structural reforms, with the other utilities being absorbed into 15 urban and 5 rural water authorities.

Having taken the decision not to privatise, there were no plans initially for independent regulation. Prior to 1 January 2004, the Essential Services Commission, responsible for the economic regulation of electricity, gas, rail, ports and grain, had a limited role in water, covering only Melbourne. Its role was limited to the development of a benchmark customer contract and customer charter, a review of performance standards, the introduction of public reporting, and the development of operational audits for compliance with licence conditions. From 1 January 2004, the State Water Industry Regulatory Order extended the role to cover all the water operations in the State (i.e. 24 utilities in total), and the functions and powers related to price regulation. This important step removed any political interference in tariff setting, with the Commission required by law to set prices to achieve full cost recovery. The utilities are required to contribute to sustainable water management through the inclusion of natural resource management costs.

The Commission's first price review set allowable water prices from July 2005 to June 2008. In its 2004 Report, the National Competition Commission said that "all prices are now set to achieve cost recovery in line with the principle of consumption based pricing". However, the discussion paper [3], (already mentioned above in the context of Brisbane), indicates that Melbourne has one of the lowest levels of cost recovery compared with other Australian cities, based on efficient replacement cost of the infrastructure, with a figure of 68%. This is compounded by State subsidies of around 40% of local authority water infrastructure capital costs.

The State of Victoria, through its reform programme, has achieved significant improvements in utility performance. Figure 7.6 shows one aspect, that of operating costs in the city of Melbourne, compared with the average for Australia. This was before the Essential Services Commission took over responsibility for tariff setting, which was intended to achieve a position of self-financing and full cost recovery. The Water Industry Regulation Order 2012 [12] makes provision for a sustainable revenue stream, and for utilities to recover their operational, maintenance and administrative costs, to recover their expenditure on renewing and rehabilitating existing assets and to recover a rate of return on investment in new assets. The regulation order provides the basis for tariff setting and planning. The utilities produce water plans to deliver the service requirements which are considered by the Commission in determining future water tariffs.

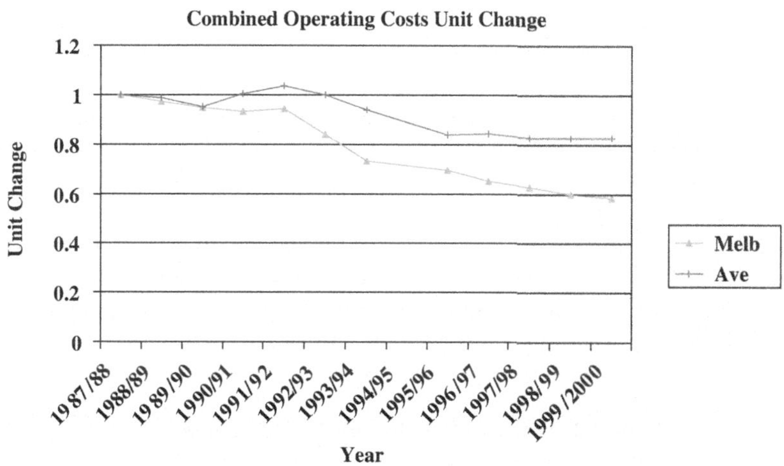

Figure 7.6 Graph showing relatively high reduction in operating costs in Melbourne compared with average for Australia over the period 1987–2000 (*Source*: John Langford presentation at International Water Association/Japan Waterworks Association/Federation of Japan Water Industry Workshop, held in Tokyo, May 2004).

It is interesting to speculate on which changes will have had the biggest impact. Will it be because of the change to larger more viable units, or because of the move to corporatised-public companies bringing private-sector practice to improve operational management; what will be the benefits of independent regulation? I believe all three to be important. Without scale it is not possible to attract good quality management, operational management should be separated from policy and political interference in day to day matters, and in this case independent regulation was required to make difficult decisions on tariffs to achieve sustainable cost recovery.

THE NETHERLANDS

The reform of the water sector began in The Netherlands in 1975 when the Government decided that water service delivery should be concentrated into larger utility operators serving a minimum of 100,000 connections. This was to create utilities large enough to attract good quality management, and to have their own quality-monitoring laboratories and technological know-how. It was stipulated that each organisation would be accountable from source to tap, and that there would be vertical integration of production and distribution. There would be no more national subsidies for unprofitable utilities, and by moving to a larger scale it would be possible to even out tariffs over larger areas. Importantly, the move would be to more managerial decision-making and fewer bureaucratic political procedures. To achieve a change in culture, municipal operations became corporatised-public companies, which in The Netherlands are called private-law companies. This requires them to operate as if they were private companies, with private corporate governance, including private sector accounting, although the shareholders are the local authorities in the areas they serve. These companies although publicly owned are not subject to governmental labour relations. Provincial plans were made for mergers and acquisitions, in which designated cities took over other cities' operations, to achieve the specified scale of a minimum of 100,000 connections. Those cities which resisted a take-over were forced to comply through court procedures, so it was very much a national policy imposed on local government. At the same time the national Government commissioned a national development plan for drinking water supply, and enhanced drinking water quality regulation, through a reinforced drinking water inspectorate. The impact of the Government's policy is illustrated in Table 7.3 below.

Table 7.3 Impact of 1975 Reform in The Netherlands.

Year	Number of Water Service Companies	Total Workforce
1980	94	8500
1990	56	8400
2000	20	6400
2004	13	5800

Source: Presentation by Piet Jonkers, Managing Director of Duinwaterbedrijf Zuid-Holland, at International Water Association/Japan Waterworks Association/Federation of Japan Water Industry Workshop, held in Tokyo, May 2004.

With further mergers the number of water service companies in 2010 was 10 with a total workforce of 5200.

The structure of each water company is shown in Figure 7.7. The companies are not-for-profit organisations, and there is no dividend payment to the local authority shareholders. The companies, which enjoy a high credit rating for investment, are governed by a two-tier board structure. The Executive Board of Directors is composed of professional managers, who have full powers of representation of their company, and have the authority to hire and fire employees. Half of the Supervisory Board of Directors are local politicians and the other half are independent experts. These boards hire and fire management and approve investment decisions. Representatives of the shareholders, the local authorities, who hold shares in proportion to their size, meet in a General Assembly. The General Assembly approves water tariffs.

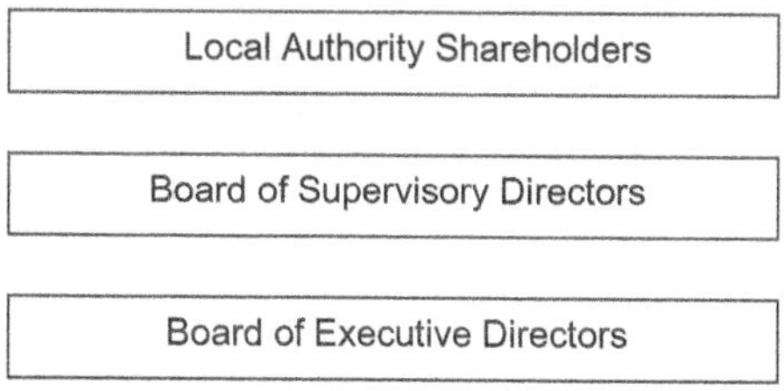

Figure 7.7 Structure of Water Companies in The Netherlands.

In 1997 the water companies established a voluntary benchmarking scheme, with most companies participating. They derived an efficiency index, based on best practice. Overall, the combined participants made improvements from an efficiency figure of 0.72 in 1997 to one of 0.88 in 2002. The companies also estimated how much cost-saving potential remained. Table 7.4 shows the figures for Duinwaterbedrijf Zuid-Holland.

Table 7.4 Figures showing cost saving potential in Duinwaterbedrijf Zuid-Holland.

Year	Cost Saving Potential € million/year			Tariff €/m3
	Definition 1	Definition 2	Definition 3	
1997	21			1.77
1998	19			1.82
1999	16			1.81
2000	10			1.78
2001	8			1.71
2002	7	11		1.69
2003		8	11	
2004			10	

Source: Presentation by Piet Jonkers, Managing Director of Duinwaterbedrijf Zuid-Holland, at International Water Association/Japan Waterworks Association/Federation of Japan Water Industry Workshop, held in Tokyo, May 2004.

The table also gives the movement in tariffs during the period 1997 to 2004 with a steady reduction from 1998 onwards. It is interesting to note that over the period, the company changed its definitions of cost saving potential twice, recognising from experience that there was greater potential for cost saving. This is to be expected with a continuous improvement in understanding, better science and advances in technology providing the potential for improvements, even though in cost savings there will be diminishing returns. Table 7.5 gives the potential-savings figures based on Definition 3 and the tariffs for the period 2003–2011.

Table 7.5 Potential Savings and Tariffs for Duinwaterbedrijf Zuid-Holland for the Years 2003–2011.

Year	Potential Savings € million/year	Tariff € /m3 /m3€ /m3
2003	11, 0	1.67
2004	9, 8	1.69
2005	14, 7	1.69
2006	7, 8	1.67
2007	7, 8	1.57
2008	7, 4	1.57
2009	6, 4	1.64
2010	3, 0	1.61
2011	3, 0	1.66

Source: Piet Jonkers, Personal Communication.

Clearly, the Dutch model has been very successful. The integration of small utilities led to companies of viable size able to attract good quality management with good scientific support services. The voluntary benchmarking scheme provided an additional incentive for management to outperform. The Government decided subsequently that, from 2007, involvement in benchmarking would be mandatory. This move created a nationally monitored system of comparative competition. There is a form of time-limited tenure, with company Managing Directors appointed by the Supervisory Board, and operating under fixed period contracts. However, there is no independent regulator (but see below some provisions in the Water Law 2011), with the success of the system dependent on each Supervisory Board exercising a regulatory role, not becoming too close to the Management Board, and not becoming too protective of their companies. The potential savings targets provide an incentive to continue to strive for improvements. In the longer term what incentives will there be to retain efficient companies? Could it be the threat of privatisation, and the motive to demonstrate that the Dutch companies can match private-sector performance?

The Water Law 2011 confirmed that tariff decision making would remain with the local authority shareholders, but stipulated some ground rules. Government would decide on the cost of capital to be used in the tariff calculations with a cap of 6% for the year 2012/13. Only "necessary" costs were to be included in calculating tariffs, although what was necessary to be decided by local authorities. The Drinking Water Inspectorate became part of a wider inspectorate, the Inspectorate for Environment and Transport, with concern amongst the water service companies that expertise on water and health matters had been diminished. The water service providers were mandated to have "permanent" consultation with the Competition Authority in line with requirements on energy and telecom companies. The longer term implications of this are unclear.

SINGAPORE

When looking at the highly developed Singapore water management of today it is important to recognise that back in 1965 the city-state was facing the same problems of water rationing and industrial pollution, whilst being relatively poor, as many other countries are facing today. The Singapore story is important because it illustrates many aspects of the ingredients of success including a clear vision, a holistic approach, long-term planning integrated into city planning, sustainable development, effective legislation and enforcement and good operational management free from corruption. The high technology photogenic aspects such as the barrage and re-use membranes get the headlines, but the "bread and butter" aspects of water distribution management, with non-revenue water down to less than 5%, are equally important and possibly more relevant to many other cities in the world. A comprehensive account of the Singapore story was published in 2013 [13].

HAIPHONG, VIETNAM

This is the story of a long period of support from the Finnish Government's Department of International Development, Finnida, to the city of Haiphong, in Vietnam. I am indebted to Eero Kontula for his permission to use the content of a presentation. The support programme had three beneficiaries, WSCO, the Water Supply Company, SADCO, the Sewerage and Drainage Company, and URENCO, the Urban Environment Company. The results of the support to WSCO are given in this account. There were four phases to the work:

- *Phase I, 1990–1993*: A technical assistance programme covering strengthening management procedures, human resources development, preparation of a water supply development plan, preparation of an organisation and manpower development plan, a financial development plan, and leak detection studies and other work to understand and improve the operation of the distribution network.

- *Phase II, 1993–1997*: Further strengthening of management procedures and human resources development, drinking water quality studies and establishment of drinking water quality monitoring, improvement of the consumer service, water distribution network analysis and the establishment of a water network telemetry system. Also, physical improvements to the water treatment works, and the distribution system, and strengthening of the network with the construction of 15 km of transmission mains.
- *Phase III, 1997–2000*: Study on billing and revenue collection, study on reduction in non-revenue water, study on protection of water resources, preparation of operation and maintenance systems and manuals, installation of control system at a treatment works, improvements to water treatment, and establishment of a GIS (Geographical Information System) mapping system.
- *Phase IV, 2000–2004*: Technical assistance to strengthen management, improve financial management, and develop operational and maintenance procedures.

The problems in Vietnam were seen as a series of vicious circles. These reflect closely Arthur McIntosh's analysis of the factors preventing access to sustainable water services in his book *Asian Water Supplies, Reaching the Urban Poor* [14]. The circles, presented earlier as Figure 3.1 but reproduced again below as Figure 7.8, are self-explanatory. The problems were categorised under two headings, inappropriate policies and poor consumer relations.

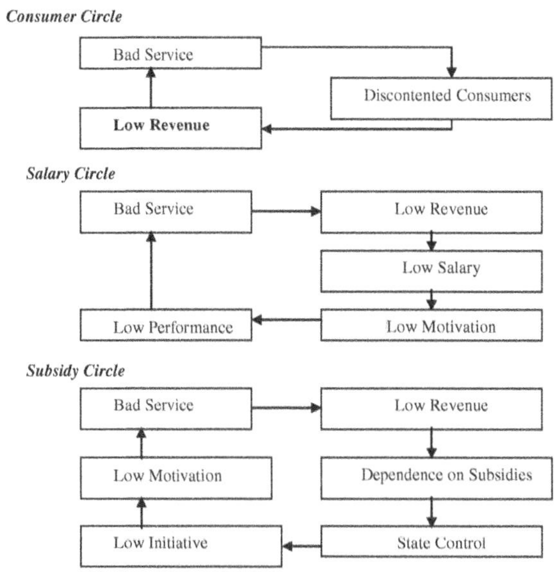

Figure 7.8 The Vicious Circles (*Source*: Presentation by Eero Kontula, Finnida at World Bank/WaterAid Workshop, London, August 2004).

Under inappropriate policies, the following were included:

- Planning and construction carried out by central Government without involvement of Haiphong Water Service Company;
- No opportunities for public participation;
- Very low cost recovery – only US$ 1 cent per cubic metre;
- No consumer metering and thus no incentive to conserve water;
- Due to top-down policy, a lack of responsibility by water company management resulting in a lack of initiatives and commitment;
- Water company management had no authority to hire and fire staff;
- No incentives for the company to improve services;
- No contract between company and consumer to define service deliverables.

The problems arising from poor policies were compounded by poor consumer relations:

- No consumer participation;
- Consumers not seen as customers having rights of redress;
- Consumers alienated by a poor level of service with intermittent supplies, poor pressure and doubtful drinking water quality;
- A low level of access to piped supplies with only 25% (80,000) of the population connected, and the remaining 80% relying on public tanks, polluted groundwater, or high cost vendor water;
- Some company employees were involved in taking bribes to make illegal connections;
- Some revenue collectors took bribes to undercharge for water.

The approach taken to reform was to begin with a mission statement by the company to provide an adequate supply of good quality water to all users, and to establish two important principles; the company should have the responsibility to plan, implement and manage the water supply system in the most effective way possible, and to finance the cost of operations and capital from user charges. In other words, the company was to achieve sustainable cost recovery. These aims were to be accomplished by winning the support of the city administration, winning the confidence of consumers through public participation, empowering and training staff to take responsibility as service providers, and introducing a strictly managed performance-based incentive scheme for staff. The impact of the changes is shown in Table 7.3.

There was further progress during the 2000–2004 Phase IV programme [15], with the company able to decentralise more management responsibility, and management information systems and financial management systems improving day-to-day operations, and contributing to long-term planning. The company now provides safe drinking water continuously, 24 hours a day. Water treatment filter performance improved through both increased production rates and also improvements in quality through achieving low turbidity treated water.

Non-revenue water was reduced from 41.3% in 2001 to 27.5% in 2004 with the prospect of further improvements subject to investment in network rehabilitation. In 2004 cost recovery was at 114% of operation and maintenance costs, so that small investments could be made from revenue income. The total charges for water supply, wastewater and solid waste services combined are reported as being less than 5% of low income families' monthly income. This would suggest that there is scope for further increases in charges to achieve the level necessary to rehabilitate the distribution systems and achieve fully self- sustainable systems.

By any consideration, the Finnida approach of technical assistance over 15 years has been very successful, (See Table 7.6) with the water company now achieving community-wide access, with 24-hour continuous supplies and the prospect of sustainable cost recovery. Finnida list the lessons learnt as starting with a carefully chosen pilot, focusing on consumers' needs and giving them a supervisory role, building employees trust, developing their skills and giving them incentives to perform well. I would add to the list the recognition at the start of the reform process of the need to achieve sustainable cost recovery, and persuading consumers that water charges would have to be increased to achieve a sustainable service.

Table 7.6 Impact of reform programme in Haiphong, Vietnam.

Performance Measure	1993	1999
House connections	80,000	380,000
Unaccounted-for Water (%)	70%	16.4%
No of meter connections	0	81,000
Employees per 1000 connections	30	7.4
Hours/year with no supply	1070	27
Households paying for water	25%	98%
Profit/Loss	−VND 3.1 billion	+VND 6.9 billion

Source: Presentation by Eero Kontula,
Finnida at World Bank/WaterAid Workshop, London, August 2004.

As to the future of the Haiphong water service, continuing success will depend upon the ongoing management operating successfully without the technical support of Finnida. The 2000–2004 report refers to the water company "being more or less ready" to become a Limited Liability Company, but refers to the process of fixing tariffs causing restrictions in financial planning and budgeting. This is limiting the company's ability to maintain the status quo, let alone to further develop its operations. This suggests the need for some independent form of regulatory body, to fill the gap left by Finnida's departure, working closely with consumers to fix tariffs based on need, and to continue the drive for improved performance and sustainable cost recovery.

The support provided by Finnida prepared the utility's management for the challenges to come. Haiphong is not only the third largest city in Vietnam but also the main port for the capital, Hanoi. It has experienced high annual growth, 11% per year compared with the national average of 8%. This rapid expansion has stretched the capacity of the water supply infrastructure. The reforms of the 1990s described above, which included financial assistance from the Asian Development Bank (ADB), had focussed on the city's five main urban districts, but had not included peri-urban or rural areas. The second tranche of the Vietnam Water Sector Development Multitranche Financing Facility, approved by ADB in December 2012, includes investment [16] in a treatment plant and distribution mains to supply Haiphong's peri-urban and rural areas. It is anticipated that there will be 90% coverage by 2020. The Haiphong situation is a good example of the impact of rapid urbanisation. See discussion on urban growth and planning in Chapter 3.

PHNOM PENH, CAMBODIA

A number of papers [17][18][19] have been written on the successful reform of the water service in Phnom Penh, the capital of Cambodia. In 1993 following the fall of the Khmer Rouge the urban water service performance in Phnom Penh was poor but by 1996 it was acclaimed as the best in class in South Asia. Some key performance figures are given in Table 7.7. Management was restructured under the highly effective leadership of Ek Sonn Chan. Staff salaries were raised with bonuses for exceptional performance, but with penalties for "unacceptable" behaviour. Any member of staff involved in corruption was sacked. There was good public consultation so that the people understood the need for charges to cover costs and were encouraged to report leakage and illegal connections. 32 out of 38 poor settlements were provided with piped water access for the first time with municipal standpipes provided for the other six. These system extensions to poor areas were funded from revenue, and there were subsidies for connection charges with the level depending on the level of poverty. It was announced in 2010 at Stockholm World Water Week, when Phnom Penh received the Stockholm Water Industry Award, that 100% water supply access coverage had been achieved. Even though access coverage had quadrupled, new water sources were not required as "additional" water had become available from a reduction in leakage. This is illustrated in Figure 7.9 but note that water sold is given as a volume increasing without a numerical scale and non-revenue water, which is decreasing, is given as a percentage. There is also ongoing provision for low-income groups as water bills can be paid in instalments and a progressive tariff with a lifeline tariff for the first band of 7 m^3 per month. This example ticks all the boxes for a successful reform programme: clear vision and sound policies with long-term planning; good management free from political interference and corruption; 24/7, asset management and attention to non-revenue water; good public consultation with provisions for low-income groups and sustainable cost recovery. To my mind the full story has yet to be told. I hope that Ek Sonn Chan will find the time to give a full personal account in a book of how it was during the dark days of Khmer Rouge right through to the exemplary water service of today. He has shown that water services for all are achievable, affordable and financially sustainable.

Table 7.7 Phnom Penh Urban Water Performance Indicators, 1993–2006.

Performance Indicator	1993	2006
Coverage Area (%)	25	90
Number of Connections ('000s)	27	147
Supply duration (hours/day)	10	24
Revenue Collection (%)	48	99.9
Non-revenue water (%)	72	8
Staff per 100,000 connections	22	4
Financial Situation	Subsidy	Full Cost Recovery

Source: Ek Sonn Chan Presentation, Stockholm WWW 2007.

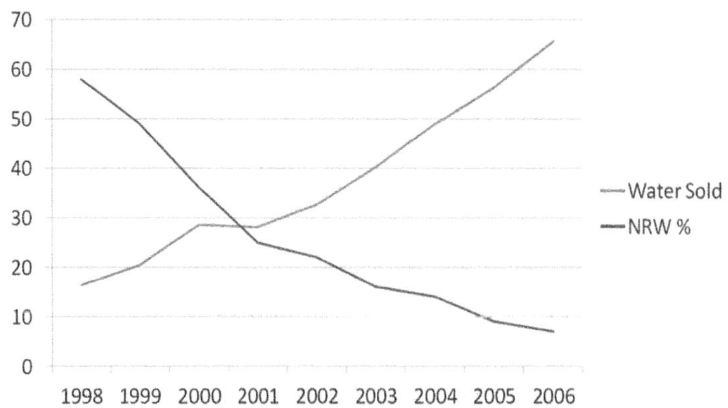

Figure 7.9 Illustration of how reduction in NRW provided increased water resources (*Source*: Author based on Vermersch *et al.* [19]).

NWSC UGANDA

A comprehensive account of the management aspects of the turnaround of the National Water and Sewerage Corporation in Uganda is given by William Mulhairwe in his book published in 2009 [20]. The problems were identified as poor customer service, poor organisational culture, large debts, low cost recovery, high non-revenue water and corruption amongst staff. A new Management Board was appointed with the task of transforming a very poor water service operation. The Board was given the freedom to operate without political interference. One key aspect of the approach taken by the new Managing Director was to delegate responsibility to area managers, to give them performance contracts and to stimulate performance competition between areas. The benchmarking and checking elements of the approach taken are discussed in Chapter 6. The improvements in performance achieved over a 13 year period are shown in Table 7.8. Although service charges were increased annually to keep pace with inflation, currency changes and key input price changes, it was not considered feasible to move to full cost recovery over that period of 13 years. Mobile phone water payment options [21] through banks and telephone service providers were introduced. This was beneficial both to NWSC through facilitated improved collection efficiency, and to consumers as it provided more convenient and less costly payment transactions. Although this reform programme has not reached the level of coverage, service and finanical sustainablility achieved in Phnom Penh, NWSC is a more complex operation as it covers a larger geographical area, and required a different approach. It is hoped that there will be the political and financial commitment to allow NWSC management to complete the reform process to achieve financial self-sufficiency and thus service sustainability.

Table 7.8 Urban Water Performance Indicators in Uganda, 1998–2011.

Performance Indicator	**1998**	**2011**
Service Coverage (% of population served)	48	75
Total piped connections	50,826	272,406
New Connections per year	3,317	25,000
Household metered Connections	37,217	271,734
Staff per 1000 Connections	36	6
Collection Efficiency	60%	98%
Non-Revenue Water	60%	33%
Metered Accounts	65%	99%
Annual Turnover (USD million)	8	51
Profit (USD million)	−3 (Loss)	12(Surplus)

Source: Mulhairwe 2009 [20].

GHANA

This final example is a reform in transition. In Ghana, the Government has created an almost ideal structure with various authorities established, with all the key functions. The overall structure is shown in Figure 7.10. MWH, the Ministry of Works and Housing, is the water policy ministry. PURC, the Public Utilities Regulatory Commission, is the economic and quality regulator. WRC is the Water Resource Commission responsible for extraction permits, EPA is the

Environment Protection Agency, and MoH, the Ministry of Health, has overall responsibility for drinking water quality. All urban water services for the whole of Ghana have been integrated under just one water service utility, GWCL, Ghana Water Company Limited, and CWSA, the Community Water Service Association, oversees rural supplies. As piped supplies do not reach all urban consumers, there is an extensive private water tanker service and also local water vendors.

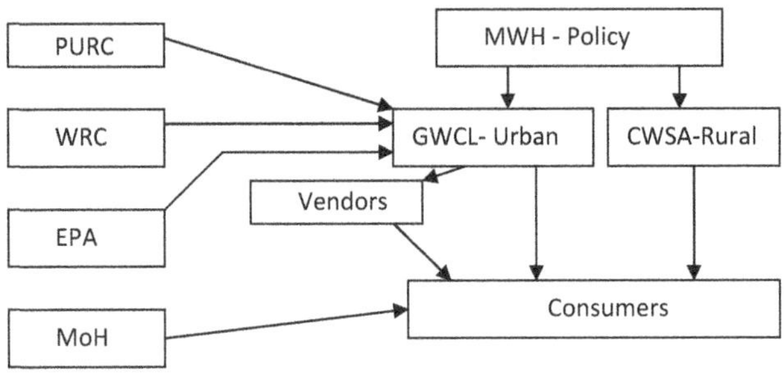

Figure 7.10 Organisation of Water Services in Ghana (*Source*: Public Utilities Regulatory Commission, PURC).

Ghana has suffered from the common malaise of under-recovery of costs with a resultant lack of maintenance and refurbishment of the water distribution infrastructure, with consequent high levels of leakage (although not measured), loss of pressure and intermittent supplies. The water treatment works are in good physical condition due to investment by donors, but often the source water is poor, due to high levels of nutrients and the formation of algae, affecting the amount of water produced. In my view, the new structure has yet to make the desired impact due to two major factors. Firstly, although PURC fixes tariffs, decisions on their implementation are made by Government, which has been reluctant to increase prices. Secondly, the management capacity of GWCL has been inadequate to meet the many challenges. There have been at least six Managing Directors during the 9 years of the life of GWCL. At the time PURC was formed in 1997, it had been the intention to "privatise" water services, but due to strong and organised resistance in the country to privatisation, it was shelved and the decision was taken to regulate a public sector operation. I sense that the Government has not seen the same need for strong regulation that it would have done in a private-sector situation. In May 2006, the management of GWCL was contracted out to a "private" company, Aqua Vittens Rand, AVRL, which was a company formed jointly by two highly respected public sector water companies, Vitens from The Netherlands and Rand Water from South Africa. Even with the public service pedigree, there appears to have been a campaign to undermine AVRL, with derogatory articles appearing in the local newspapers.

The 5 year contract was not renewed. One period of five years was not long enough to achieve effective reform, especially as the contractor had no control over investment in operational development. The type of management experience the contractor was bringing to Ghana water is necessary to tackle the fundamental problems. Ghana has suffered from the vicious circle ailments of Figure 7.5. Following the decision not to renew the AVRL contract a subsidiary company of GWCL, Ghana Urban Water Limited (GUWL) was formed to manage operations. The first MD of GUWL had been the chief operating officer of AVRL. I do wonder why it is felt necessary to retain a holding company with just one operating subsidiary. It is to be hoped that GUWL will receive management and investment support from GWCL in tackling the major problems being faced in urban water services. Political and financial commitment to achieve 24/7 is necessary without which it will not be possible to extend a sustainable service to all. Having a good governance structure is insufficient. As can be seen from the successes in Phnom Penh and NWSC Uganda, good management has to be allowed to operate without political interference. It is the poor who are suffering from inadequate water distribution because many of them in Ghana are currently paying water vendors very high prices for water, with some estimates as high as 14 times that for the piped supplies. Once again the subsidies are benefiting the more wealthy members of society, and holding back the necessary reform to allow the development of sustainable water services for all. It is instructive to compare the current position in Ghana with the high quality service enjoyed by all in the city of Haiphong in Vietnam.

There is one area where Ghana could have provided a model for the world, and that is in the "management" of water tankering operations. As already mentioned, due to poor coverage of water supplies to the extremities of the capital city Accra, many people rely on road tankers. The tanker operators have formed themselves into four associations as a means of collectively negotiating with GWCL on the availability of good suitably located filling points, and on the price of water. PURC has worked with these associations to produce Tankering Guidelines [22], which are a code of practice to be self-regulated by the associations and audited by PURC. The guidelines include a requirement that the tanks are disinfected at least every six months, and that there should be random checks on water quality in the tankers. The tankers are required to carry a registration mark on the side of the vehicles, and consumers can check the price they should be charged with the tankering associations. This appeared to be working well but sadly in 2012 [23] following complaints of poor pressure from some consumers connected to the distribution network, GWCL responded to political pressure and closed most of the tanker filling points. If there had been good network analysis of the distribution networks the reasons for the poor pressure could have been determined. It is possible it resulted from deterioration of the distribution system. If it had been proven to be as a result of tanker abstraction, more suitable filling points could have been identified. Whatever the reason, it is the poor who have suffered. I believe the tankering guideline approach to be sound. Tankers will also be necessary to reach the un-served until the distribution systems have been refurbished and extended.

GENERAL LESSONS

The various examples described above cover both success stories, and work in progress. They all demonstrate the need for separation between policy and operations, and between government and water service provider, with the latter being able to manage without political interference. The Seattle story shows that reform can take place very successfully within a traditional municipal structure. This is also true of Brisbane to a certain extent, although the situation there is marred by the water utility being used to provide income for other city services. However, generally the route to reform has been the integration of small utilities to create those of a size able to attract good quality management, and a change to corporatised-public companies. Such companies are required to operate as if they were private companies, but remaining in public ownership. The Netherlands and the State of Victoria in Australia are good examples, with The Netherlands driving reform through benchmarking and Victoria deciding that an independent regulator is required to achieve full cost recovery. Scotland, in the UK, is another good example, where a large public-corporatised company has achieved significant improvements in performance, driven by a regulatory system equivalent to that for the private-sector situation in England. In Chapter 8, I describe how Scottish Water has successfully used partnering to deliver a very large infrastructure construction programme. The story of Haiphong shows what can be achieved through management and staff development, but also that it takes a number of years. The common thread throughout is the need for full (sustainable) cost recovery.

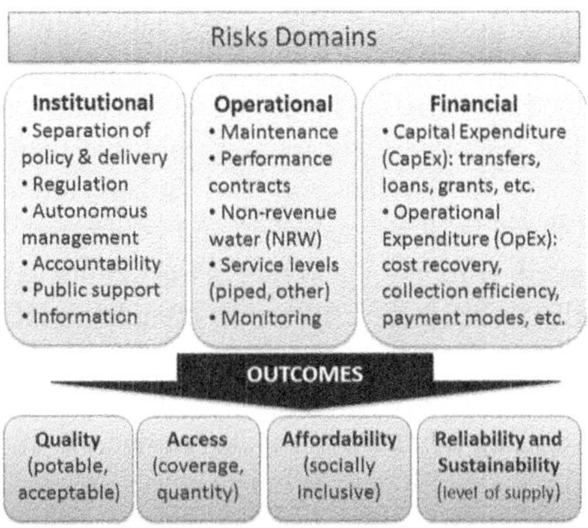

Figure 7.11 Risk Domains (*Source*: Hope and Rouse [24]).

Reform can be considered as managing risk. Hope and Rouse [24] considered risks in three domains, those of institutional, operational and financial. The diagram, Figure 7.11, taken from the paper gives the factors and outcomes. Those elements have generally been managed successfully in the examples given in this chapter.

REFERENCES

1. National Competition Council Annual Report 2005–2006, Oct 2006 www.ncc.gov.au (accessed July 2013)
2. WSAA Facts. www.wsaa.asn.au (accessed July 2013)
3. Securing Australia's Urban Water Supplies: Opportunities and Impediments (2006). A discussion paper prepared for the Department of the Prime Minister and Cabinet. Marsden Jacob Associates, November 2006.
4. South East Queensland Water (Restructuring) Act 2007. State of Queensland 2007. http://www.legislation.qld.gov.au/LEGISLTN/ACTS/2007/07AC058.pdf (accessed July 2013)
5. South East Queensland Water (Restructuring) and Other Legislation Amendment Act 2012. State of Queensland 2012. http://www.legislation.qld.gov.au/LEGISLTN/ACTS/2012/12AC039.pdf (accessed July 2013)
6. Report of the Walkerton Enquiry. Queen's Printer for Ontario. ISBN: 9780779425594.
7. Safe Drinking Water Act 2002, Ontario Ministry of the Environment. http://www.e-laws.gov.on.ca/html/regs/english/elaws_regs_030170_e.htm (accessed July 2013)
8. Annual Report 2004–5 Chief Drinking Water Inspector Ontario Ministry of the Environment, April 2006, Queen's Printer for Ontario. ISBN: 9781424931972.
9. Watertight: The case for change in Ontario's water and wastewater sector. Queen's Printer for Ontario, ISBN: 0779479289. http://pubont.stores.gov.on.ca/pool (accessed July 2013)
10. The Water Opportunities Act 2010: Ministry of the Environment, Ontario, Canada.
11. 2012 Annual Report: Water Technology Acceleration Project (WaterTAP). www.watertapontario.com (accessed July 2013)
12. The Water Industry Regulation Order 2012: Essential Services Commission, State of Victoria, Australia, www.esc.vic.gov.au/getattachment/Water/Water-Price-Review-2013-18/Wiro-2012-wa-20121018-(1).pdf.aspx (accessed July 2013)
13. Tortajada C., Joshi Y. and Biswas A. K. (2013). The Singapore Water Story: Sustainable Development in an Urban City State. Routledge, London.
14. McIntosh A. C. (2003). Asian Water Supplies: Reaching the Urban Poor. Asian Development Bank, ISBN: 9715613802.
15. Water Supply (2004). Drainage, Sewerage and Sanitation Management Programme in Haiphong 2001–2004 Project Completion Report, Finnida, December 2004.
16. Rehabilitating and Upgrading Project of Haiphong Water Supply System: Asian Development Bank 2011. www.adb.org/projects/documents/water-sector-investment-program-draft-facility-administration-manual (accessed July 2013)
17. Biswas A. K. and Tortajada C. (2010). Water supply of Phnom Penh: an example of good governance. *International Journal of Water Resources Development*, **26**(2), 157–172.

18. Chan E. S. (2009). Bringing safe water to Phnom Penh's city. *International Journal of Water Resources Development*, **24**(4), 597–609.
19. Vermersch M., Chan E. S. and Vaughan P. (2012). Are we going to win the war against NRW? *Water 21*, **14**(4), 57–59.
20. Muhairwe W. T. (2009). Making Public Enterprises Work; From Despair to Promise: A Turn *around* Account. IWA, London.
21. Foster T., Hope R., Thomas M., Cohen I., Krolikowski A. and Nyaga C. (2012). Impacts and implications of mobile water payments in East Africa. *Water International*, **37**(7), 788–804.
22. Water Tanker Service Guidelines for the Utility and Operators of Tanker Services (2007). Public Utilities Regulatory Commission, Ghana, 2007.
23. Hansen K. and Rouse M. J. (2013, in preparation).
24. Hope R. A. and Rouse M. J. (2013, in press). Risk and Responses to Universal Drinking Water Security. Philosophical Transactions of the Royal Society A, London.

Chapter 8

Steps for the successful use of the private sector

Tapping commercial energy

Whatever the form of water service provision, the private sector is always involved. Virtually all construction, whether it is for treatment works or underground pipe systems, is carried out by the private sector. The private sector provides virtually all of the hardware and consumables, and most of the software. Private sector consultants provide advice, develop schemes and carry out detailed design work. Private contractors carry out system maintenance, either in total or as a supplement to a water service utility's own workforce. In doing all this work, the private companies involved either make a profit, or they go out of business. In spite of a recognised profit element, it is accepted by those without bias, that the above mentioned work is more efficiently carried out by the private sector and that it results in benefits to water service customers. So why is the use of the private sector for financing and managing water services so controversial? Why do "failures" of private sector contracts receive extensive publicity and critical academic review? Why is it that poor performance in public sector operations, often in the same areas prior to the attempt to use the private sector, is generally ignored? This chapter attempts to look behind the emotive arguments, and to give advice based on an objective assessment of successes and failures. I begin with some examples of the history of the private water sector.

SOME EXAMPLES OF THE HISTORY OF THE PRIVATE WATER SECTOR

In London, there was private sector investment in water supplies as early as the 13th Century when French merchants, who wanted to be involved in the London wool markets, financed the Great Conduit [1]. This conduit, made of lead, conveyed water from what was the town of Tyburn, west of London, now Marble Arch, to the city area. Most people were then supplied by private water-bearers [2], who carried the water in 3 gallon or larger containers called tankards. The water collected from the conduit was free, but the water-bearers charged for their delivery service. There are various estimates of the number of water carriers, one as high as 4000, who formed an association known as the Fraternity to protect their interests. In later years, further conduits were built and from the 17th Century onwards, with the availability of steam power, there were a number of private initiatives to take water from the River Thames and deliver it through pipes to properties. The 30-mile long New River canal, taking water to London from Hertfordshire, was financed by a Welsh jeweller, Hugh Middleton. Over time, the responsibly for water supplies was taken over by the city and other local authorities, eventually all becoming operated by the Metropolitan Water Board, remaining public until privatisation of the Regional Water Authorities in 1989.

In the early days, permission to form a company to provide a water service required an act of parliament. Such permission was given to a man called William Yarnold in 1699 to supply water to Newcastle upon Tyne in the north of England. This company later became Newcastle and Gateshead Water Company, which was one of the 26 statutory private water companies included in the changes, when the Regional Water Authorities were privatised in England and Wales in 1989. Later, Newcastle And Gateshead Water Company became part of Northumbrian Water which exists today. Many of the statutory private water companies have a long history. These include Bristol Waterworks formed in 1846, Cambridge Water formed in 1853, Folkestone Waterworks Company (now Folkestone and Dover) formed in 1848, and South Staffordshire Water formed in 1853. Until 1989 these statutory water companies were regulated through a profit cap of 5%. So in England, water supply services owned and operated by the private sector are long established. They have a good reputation for integrity and service.

In France, there is also a long history [3] of the use of the private sector for the operation of water supplies. In the 18th Century private companies were granted concessions to take water from the River Seine and deliver the water in buckets to the citizens of Paris. In the 19th Century private companies began to build and operate water supply networks. The two large private companies, now known as Veolia and Suez, were formed around 1870. Today, about half of the water supplies in France are delivered by the private sector. The various forms of contract are discussed later in this chapter.

In The Netherlands, there is reference to Dune Water Company being formed in 1851 with English capital. Water was collected from the dunes and transported to Amsterdam by a large pipeline. In the book *Water of Vienna* [4] there is mention of "a number of private pipelines built in various parts of the city". It can be assumed that there were similar private initiatives throughout Europe during the early period of piped water supply.

The history in the United States of America is similar. One example is San Antonio [5] in Texas where in 1877 the city entered into a water supply contract with a private company. The company changed ownership a number of times until 1925 when the city raised money through a bond issue and purchased the system.

THE EMOTIVE ARGUMENTS

There are probably more papers written either advocating the use of the private sector, or condemning it, that on any other water subject. The number of papers condemning it probably outnumbers those supporting it by ten to one. A highly organised anti-privatisation lobby tried to hijack the World Water Forum in Kyoto in 2003. This had the unfortunate effect of attracting media attention, and overshadowing the more important issues on how to meet the water and sanitation needs of the unserved or underserved parts of the world. Given the history described above, why is the use of the private sector so controversial?

The main argument is that "water is a human right and should not be for profit". If that argument were to be taken to the ultimate degree, anything to do with water supply, including the manufacture of pipes and equipment, would have to be public sector. Equally, should the same argument apply to basic foods and basic shelter?

A more compelling argument is that since water supply is a monopoly, due to the high cost of the distribution infrastructure, it is not possible to have effective competition with the market controlling water pricing. It is largely true that direct competition is not possible, although there are elements of competition for larger customers using common carriage in the UK, and there are proposals to extend this to all non-household customers. However, an effective form of competition can be introduced through regulation-driven comparative competition.

There are also emotive arguments against the public sector, as some people believe that the public sector is inherently inefficient, as there are no market pressures to provide the motivation for management to strive for improvements. The World Bank was criticised ([6] and [7]) for adopting this attitude, generally by those opposed to privatisation. In my view, the Bank's policy needed to be challenged, but not for the reason it was being criticised. There was a need for the Bank to recognise that "one size doesn't fit all", that private sector involvement can provide substantial benefits in a well-regulated environment but that, in some other conditions, it is more appropriate to give emphasis to public utility reform. The Bank now puts a lot of effort into the capacity building of the public sector, although often this change of policy is not recognised by its critics.

I believe that all these emotive arguments are counter-productive, as there are enough examples to show that both the private sector and the public sector can be very successful in delivering good quality and cost-effective services. It is instructive to consider cases where the private sector has been criticised to help define what is necessary for its successful use.

HIGHLY PUBLICISED EXAMPLES OF THE USE OF THE PRIVATE SECTOR

I take and analyse some of the most publicised examples: Atlanta in the USA, La Paz and Cochabamba in Bolivia, Adelaide in Australia, and Jakarta in Indonesia. Although I try to get behind the emotive arguments, I have quoted from some of the newspaper articles at the time to give the local atmosphere in which the action took place. Working on the basis that all involved were trying to do the right thing, I do not attribute blame for any deficiencies, but try to draw out the lessons to be learnt from these experiences.

Atlanta, USA

The background to the Atlanta contract was a water service which had needed to expand rapidly when the population tripled during the period 1960 to 2000. Not all of the population was served under the concession contract, but at that time the city had an $82 million budget deficit. There was general acceptance that, year after year, funding for effective maintenance of the distribution system had been deferred. The sewerage system required even more attention, with an estimated investment requirement of $3 billion. Although not discussed publicly at the time, it is considered that one of the motivations for using the private sector was to reduce the cost of operating the water supply system, to provide money for investment in the sewerage system. The city placed a 20 year contract with United Water Services beginning in 1999 for the management of water supply, system maintenance, billing and consumer service. The annual fee for this work was set at $21million, in comparison with an original annual operating cost of $48 million. So the estimated saving to the city was around $400 million over 20 years. The City press release [8] in November 1998 was upbeat and full of promise. It looked to be a good deal for the city but by the end of 2002 problems began to emerge. There were articles in the local newspaper critical of the performance of the operator [9]. There were complaints from residents about "brown water through their taps with bits of debris". It was reported that the city's transport department threatened to repair a leak and send the bill to the Water Company, claiming that it had been leaking for two years, creating a "cavernous hole". It is possible that there had been brown (rusty) water problems before 1999 but there is no record of any. The result of this discontent was that a new mayor gave the company 90 days to improve the service or the contract would be terminated. The contract was terminated in January 2003, at which time the company was in dispute with the city for $80 million of "uncontracted work". The company claimed that there were signed letters with the previous administration, which had agreed to an annual increase in the contract price of $4 million per year. The situation at the time of the termination of the contract was that:

- The company had achieved cost savings of $10 million per year (half of the anticipated savings).

- However, there was a city franchise fee of $9.8 million. The implication of this was that the savings were absorbed into the City's general finance, not leaving any money for investment in the water infrastructure.
- There was a better understanding of the condition of the supply system. It was much worse than allowed for in the contract (see Table 8.1).

Table 8.1 Data showing the poor information on which the Atlanta contract was based.

Item	Basis of contract	Found in practice
Faulty water meters	1171	11108
Mains breaks	101	279
Faulty fire hydrants	734	1633

Source: Article "What went wrong in Atlanta" www.waterindustry.org.

Not all observers agreed with the decision to terminate the contract. An article in the Atlanta Journal [10] referred to United Water's service not being perfect, but "when compared with the city's past performance, United did a serviceable job". The article warned of a likely significant increase in future water rates.

In my view the contract failed because it was not designed to deal with the real challenge facing Atlanta. It did not make provision for increased investment in the water infrastructure, to make up for earlier inadequate funding of maintenance and refurbishment of the distribution system. The contract was based on poor information on the condition of the system, as shown by Table 8.1.

Clearly, there was also a need for improved operational management to deliver greater efficiency of operations. The Drinking Water Bureau, which took over responsibility for operations after the termination of the contract, inherited a slimmer organisation with 46% fewer staff than employed prior to the contract. Instead of the efficiencies being seen as providing general investment capital for the city, it would have been more realistic to see them as reducing the required level of additional investment. The improved performance would have reduced the necessary water tariff increases to pay for the required capital investment. Contrast this with the increase in investment in the UK infrastructure, funded partly by efficiency savings and partly by increases in water charges.

There are two main lessons from the Atlanta contract. Firstly, contracts need to be based on accurate information or, if good data is not available, there should be a means of adjusting the contract when the true situation is known. Secondly, the use of the private sector is not a substitute for facing up to the need for higher investment in infrastructure. Had there been a better understanding of the investment needs, the private sector could have made an effective contribution to meeting the objectives of the city. I believe also that there should be independent scrutiny of contracts, both to avoid important aspects such as system condition being "fudged" (glossed over), and to give greater objectivity and transparency in the operation of the contract, especially if difficulties arise. I discuss effective procurement later in this chapter. The conditions to achieve it are among key recommendations made in the summary in Chapter 10.

It is instructive to look at how Atlanta has addressed the infrastructure investment requirements following the termination of the contract. The new publicly-based operation in Atlanta, under the Department of Watershed Management, has put in place a major (around US$4 billion) investment programme, called Clean Water Atlanta. Although largely aimed at the needs of the wastewater and environment programme, it included some provision to replace old water mains. This programme had to be financed and without subsidies it was estimated that water service charges, covering both water and sewerage, would increase by 10% each year for 10 years.

The investment monies for Clean Water Atlanta were raised by a combination of routes; State Loans $500 million), issue of Water Bonds ($900 million) and what was described as a "line of credit" ($1.2 billion). Approved through a referendum in 2004, there was a special sales tax which over four years raised $330million. This tax was renewed in 2008 and 2012. To pay for this investment tariffs were increased year on year. See Table 8.2 for figures for the period 2007–8 to 2011–12.

Table 8.2 Increase in monthly bills over 5 years.

Year	2007–8	2008–9	2009–10	2010–11	2011–12
Average Monthly Bill in US $	84	107	120	135	152
Percentage Increase over Previous Year	–	27.3	12.1	12.5	12.6

Source: Atlanta City website.

Significant progress was reported in the 2005 Performance Report [11]. In 2008 it was announced on the website that the Clean Water Atlanta programme was half completed and gave a completion year of 2014. An audit report of 2009 [12] suggests that further investment would be required in refurbishing the distribution system. The report gave non-revenue as 29%. Burst water main numbers over the period 2005–8 quoted in the audit report are given in Figure 8.1. These numbers are higher than at the termination of the contract in 2003.

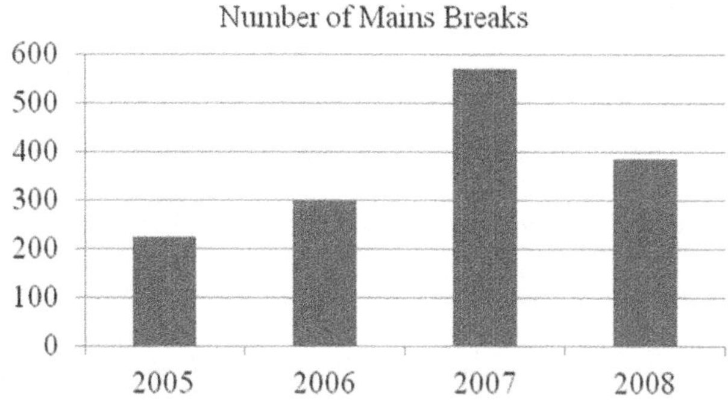

Figure 8.1 Atlanta mains breaks (*Source*: Audit Report 2009 [12]).

The Department of Watershed Management published a detailed response [13] to the audit report and advised that three distribution improvement programmes had been put in place:

- Valve and Hydrant Program (2009–2012) – locates, maintains, repairs or replaces distribution valves in the drinking water system;
- Leak Detection Program (2010–2013) – identifies non-surfacing leaks for repairs; and
- Flushing Program (2011–2014) – removes sediment from water pipes by discharging water through hydrants.

The distribution programmes listed are designed to address leakage and potential rusty water problems. The question remains as to whether further replacement/refurbishment of the distribution system, which was a key problem during the private contract, will be necessary. The City of Atlanta has faced up to the requirement for investment in infrastructure and to the unpopular need for higher tariffs to pay for a sustainable service. There has been approval for an extension of the sewerage investment programme up to 2027. The water charges in 2011 were reported to be the highest of similar cities in the United States. As might be expected there have been questions as to why. The 2009 audit queried the financial model used to determine tariff increases. The model was reviewed [14] by the Audit Office in 2012. The review concluded that while the current model projects annual rate increases of between 2% and 4% after 2016, smaller rate increases may be justified. This may be so, but the differences seem to be at the margin and perhaps the City of Atlanta should be asking the other cities why they aren't addressing their infrastructure problems; at some time they will all have to face up to major refurbishment programmes and associated increases in water tariffs. In the American Water Works Association (AWWA) report 2012 "Buried No Longer" [15] the situation is highlighted with an estimated infrastructure financing cost of US$1 trillion across the country by 2035. Atlanta (and Seattle see Chapter 7) is leading the way.

La Paz and Cochabamba, Bolivia

In the late 1990s, it was Government policy in Bolivia to require cities to privatise their utility services. This was not popular amongst the population in which a strong anti-privatisation movement developed, called the Defence of Water and Life, involving trade unions, peasant groups including Quechua Indians, environmental groups, neighbourhood associations and local Authorities. The critics of Government policy pointed to a very successful water supply operation in Santa Cruz [16] run by a cooperative, which was performing much better than the traditional local authority operations in La Paz and Cochabamba. I believe that setting up the Santa Cruz cooperative achieved some of the necessary reforms, such as separation of policy and delivery. Table 8.3 shows some of the key performance measure figures.

Table 8.3 Relative performance of three major cities in Bolivia.

	The three cities in 1997		
Performance measure	**Santa Cruz**	**La Paz**	**Cochabamba**
Unaccounted-for water (%)	23	33	54
Metered connections (%)	100	98	84
Access coverage (%)	80	84	57
Average supply hours per day	24	Not available	5

Source: Andrew Nickson paper [16].

Santa Cruz was also praised in a World Bank Report on its effective use of foreign loan finance, in which it completed the programme on time and below budget, enabling it to fund an additional project on sewerage infrastructure. In comparison, La Paz required an additional 18 months to complete its project, and Cochabamba had completed only 77% of its original programme by that time. It was necessary to improve the management of the La Paz and Cochabamba operations, and the Government looked to the private sector. I cover the two contracts together, as although they are very different in nature, they both took place within the same challenging political situation in Bolivia. The La Paz contract was initially seen as very successful but later came under criticism, whereas the Cochabamba contract had problems throughout its short life.

The La Paz contract was a 25 year concession designed to expand the service to the poor. The objectives of the contract were "unusual" (World Bank) with the emphasis on new connections. Instead of bidders being required to state the tariff they would require to achieve specific outputs including new connections, the bidders for the contract in 1997 were required to state how many additional connections they would make for a specified tariff. The contract was placed with the Aguas del Illimani consortium of which the major partner was Suez Environnement. The contract was overseen by Superinterdencia de Aquas, a reasonably independent body formed in 1994 to grant and regulate concessions for water and sanitation services. The contract included the requirement for the installation of around 72,000 new connections in El Alto, a poor neighbourhood area of the capital city, La Paz, by 2001. The Government set a tariff increase of 35% overall, but this was within a tariff block structure in which there was a low social tariff for the first 30 m^3 per month. The Government, as part of the contract conditions, set a maximum connection charge of US$155.

The Consortium, working with the Water and Sanitation Program of the World Bank, with SIDA (the Swedish International Development Agency), with the United Nations Development Programme, and with local neighbourhood associations, started with a pilot project. This project was set up to develop the implementation approach, to work through problems in making new connections, and to work with the communities in setting up payment facilities for the low-income consumers. One decision resulting from this was the option for low-income consumers to pay for the connection charge in $5 monthly instalments. This figure was equivalent to what non-connected consumers were paying for water delivered by tanker.

By the end of 2001, the contract had delivered 52,000 connections, which the Consortium considered to be full coverage, and claimed that the original 72,000 estimate was too high. In 2001 the Consortium was asked to expand the work in the contract to extend connections to communities outside the original area covered by the contract. Originally, the agreement was for 15,000 additional connections, but there were later amendments reducing that figure to 8000 connections and finally to none at all. Around this time, Government took the decision to raise the combined connection charge for water and sewerage to $450 which is half the annual minimum wage, although in some areas with a lower cost condominial system, the connection charge was about half of this. The condominial system, in which pipes are laid under paths and across properties, rather than under roads, using local labour, was developed as part of the El Alto pilot project. The Consortium claimed [17] that it had tried to persuade the Government to reduce the connection charge, but the Government had refused. The Consortium pointed out that it did not make sense for the poor to pay the same connection charge as for instance a hotel, and that there should be targeted subsidies for the poor. The public perception [18] was that it was the Consortium which fixed connection and other charges, and was exploiting the poor.

In spite of the initial success of the contract with improved service and infrastructure, the Government announced in January 2006 that it was terminating the contract. This followed opposition to foreign control of water and natural gas resulting in civil unrest that brought down the President, Carlos Mesa. He was replaced by Evo Morales who, as one of his first proposals, announced that he would terminate the La Paz/El Alto contract. In a ceremony at the presidential palace in La Paz in January 2007, Morales officially bid farewell to the affiliate of French transnational Suez "Water cannot be turned over to private business," Morales said. "It must remain a basic service, with participation of the state so that water service can be provided almost for free." Although the intention to terminate this contract resulted from a country-wide political movement, to a certain extent the anti-privatisation climate was fed by the situation in Cochabamba, where a contract survived for only 6 months.

At the time of placing the contract in Cochabamba, there were major problems. There were insufficient water sources. The water supply system was in poor condition with unaccounted-for water at over 50%. There were intermittent supplies, with an average of only 5 hours per day for those 57% of consumers fortunate to have a supply at all. The quality of water delivered for that short period of time was described as "unhealthy"; this is typical of interrupted supplies due to the likelihood of the ingress of contaminants when the system loses pressure. Prior to the contract, there was a falling block tariff in which consumers paid less per cubic metre for each block. This gave no incentive for the larger and more-wealthy users to reduce their consumption of water. All of the forementioned meant that the Aguas del Tunari Consortium, of which 55% was owned by International Water, and of which Bechtel had a 50% stake, inherited an operation in very poor condition. The contract was to operate both the water and sewage systems for the urban area only; it did not include rural supplies. Nor did it involve either ownership or leasing of water sources as had been reported [19]. Ownership of water resources remained with the Government. At the beginning of the contract, tariffs were increased by Government by an average of 35%. There was a need to finance effective maintenance of the system to reduce leakage but, according to Bechtel [20], half of the increase was required by Government to be set against over $30 million of debt left by the previous public sector operator. The finance and cost recovery aspects were compounded by the contractor being required to assist in financing a new dam. An article in the Economist [21] in 2003 reflecting on the failed contract included the statement "The mayor and the Bolivian government were wrong to insist on an expensive and unnecessary dam. But the bigger problem was that water tariffs had been too low for too long, starving the system of investment. Had the tariffs been raised earlier, more cash would have been available to improve service. These twin failings meant that any new contract, public or private, was bound to lead to unacceptable price rises". Prior to the contract, local Quechua Indians had enjoyed a free concessionary supply, but after the placement of the contract they were required to pay for their water supply. Water bills also increased due to higher consumption. Very early into the contract, through improvements in treatment plant performance, there was an increase of 30% in the amount of water going into supply. This increased the number of hours a day when water was available; consumers took more water, and bills rose accordingly. None of this was known to consumers, as there was no consultation prior to letting the contract. The anti-privatisation movement was able to blame the private consortium for the price changes. There were false reports of some consumers paying three times as much. The contract had been let in November 1999, and protests started in February 2000. These became violent with thousands of protestors met by the police and the army using tear gas. The protests extended to other parts of Bolivia and six people died. Immediately, the Government reduced the water tariffs back to their pre-contract levels. The Government rescinded the contract in April 2000; it had lasted for just six months. There followed a contract dispute which was resolved in 2006 with no payments being made. Jorge Cuba, a Bolivian journalist writing in March 2002 [19], said "the departure of these foreign firms is not necessarily the best solution, not even for the poor". There is a need for both investment and good management in Cochabamba, and it is too early to know whether the water services situation will improve in this city, and elsewhere in Bolivia, under the new President's policies.

So what can we learn from these two cases? I see four aspects:

- Firstly, public consultation and input are critical. In La Paz/El Alto this resulted in a successful early period of that contract. In Cochabamba, this seemed to be missing altogether with a disastrous result. Even in La Paz insufficient effort was made to explain the need for investment, and thus the need for increases in water charges.
- Secondly, any contract is doomed to failure if tariffs have to be raised too quickly to address lack of investment, inadequate cost recovery and associated poor maintenance. Post contract termination the fundamental problems remained; the situation in 2013 is discussed in the concluding paragraph.
- Thirdly, sufficient attention should be given to the needs of the poor. In El Paz the impact of the initial connection charge on the poor of $155 was alleviated by the $5 a month payment option, but there should have been targeted subsidies to help the poor when the connection charge was increased by the Government to $455. This was an unacceptable connection charge. Even if it could be argued that, in the long term, the poor would be better off than when they were buying very expensive tankered water, the important difference is that for tankered water consumers "pay as they go". The poor live "day by day" and are unable to save money for big one-off payments.
- Fourthly the private sector should not be used in a political and social climate that is hostile to its involvement. This is particularly the case where the required institutional and regulatory frameworks either do not exist, or there is a lack of experience in managing the private sector. Both La Paz/El Alto and Cochabamba are likely to have benefited from the improved management input during the period of the contracts, and it is hoped that the staff still involved will be able to apply the lessons they will

> have learned. Whether public or private, there is a need for sound policies, sufficient investment, good consumer consultation and information, and good operational management free from political interference.

So what are the situations in Le Paz and Cochabamba following the termination of the contracts? There are severe water resource problems due to retreating glaciers and increasing groundwater access depths. In Cochabamba it was reported in 2011 [22], eleven years on, that access had not improved. Still only 48% of the population was connected to the water supply network with sanitation coverage being about the same. The management of SEMAPA, the public utility having taken back responsibility for water services, complained [23] that there is insufficient money for development, they are unable to raise tariffs and are unable to obtain external capital. With SEMAPA unable to extend the service the remainder of the population either is paying a much higher price (up to 10 times [24] those receiving subsidised piped water) to water vendors or is served by small community constructed and managed systems. One example [25] is a Community Association for Drinking Water and Sanitation which was formed on April 22, 2007, as a public service community association with charitable status. The neighbourhood organised itself to manage the installation of a distribution network. Two tanks were installed, and a road was built to provide for tanker access to fill them. It created a Steering Committee to oversee the work. The water association had 748 members in 2011 and finances itself from tariffs and donations. The storage capacity of 250,000 litres, although better than individual purchase from water vendors, is a small amount to serve that population. But such organisations are not alone. Since 2004, many of these community-run water systems have come together to form the Association of Community Water Systems of the South (ASICA-Sur) [26]. ASICA-Sur has obtained funding from the European Union to build independent water systems. These independent systems buy water in bulk from the SEMAPA but are managed by the users. Will they be a sustainable solution for the unserved in Cochabamba? Community managed systems have the benefit of local pride and ownership (See Chapter 5) but are generally quite small. Scale can be achieved through a co-operative approach as with the SISAR systems in Brazil. Will the service to those served directly by SAMAPA improve? Will there be sufficient income to maintain those systems and achieve 24/7? There are no easy solutions. The question is not whether it should be private, public or community managed systems, but rather having policies, planning, financing and management which address the fundamentals of water service provision.

The situation in La Paz was better than Cochabamba both before and during the contract period. What happened after 2006? Over the period 2007–10 Epsas, the state-owned utility, continued the process of new connections [27] but at a rate of 36% less for water supply and 29% less per year than achieved during the contract period. It was announced [28] in February 2013 that a project, financed by the Facility for Infrastructure Development (ORIO), a grant scheme operated by the Dutch Ministry of Foreign Affairs, would bring improved water services to 125,000 people in a peri-urban area of La Paz by 2015. The article expressed concern over the project's sustainability unless there were to be a national sector strategy on how to finance the full costs of service delivery. At the time of the announcement tariffs covered only direct operations and some maintenance costs. To assist the poor there was a "solidarity" tariff for the first 10 cubic metres of water supplied but it was unclear how this tariff would be subsidised. There was reference to performance indicators on service levels provided, cost recovery, investments, customer satisfaction, and operation and maintenance (O&M). It would be good if this project were able to inject some realism into the provision of sustainable water services in Bolivia. Whether public or private, maintenance and ongoing refurbishment of systems are essential for sustainability, and this cannot happen without sustainable cost recovery. (See discussion on cost recovery in Chapter 3)

Adelaide, Australia

A process of reform [29] of Australia's water sector began in 1994 when the Council of Australian Governments established a strategic framework. This framework covered economic, environmental and social objectives, and included:

- pricing reform with a move to consumption-based pricing and full-cost recovery;
- a requirement to give transparency on subsidies and to eliminate cross-subsidies with other services;
- a requirement for there to be an allocation of water to the environment to address the low flow river problems; and
- a requirement for public consultation and participation.

Three dates were set for the states to meet these requirements:

- 1998 for urban water pricing and institutional reform;
- 2001 for rural water pricing and environmental water allocations to stressed rivers;
- 2005 for environmental water allocations to all river systems and groundwater.

In April 1995, this water reform programme was taken under the umbrella of National Competition Policy and the jurisdiction of the National Competition Council (NCC). The National Competition Policy covered all aspects of state and local authority operations and was aimed at making local services both more effective and less costly. The process of reform was overseen by the NCC, which carried out annual assessments of Governments' compliance on all aspects of the reform requirements. As an incentive to the states to meet the reform requirements, there was a per capita payment by the Commonwealth Government to the states. This payment was made when the NCC was satisfied that the reform conditions had been achieved. Generally, the states recognised that it would not be possible to achieve the necessary reforms with traditional municipal operations. The deficiencies of the traditional approach have been discussed earlier in Chapter 7; they include the non-viability of many small operations and the lack of separation between policy, regulation and delivery. However, Brisbane (see Chapter 7) found a way of achieving separation by means of internal contracts. Each state could choose how it met the reform requirements, and whether or not it would use the private sector. The policy in the State of Victoria is discussed under the corporatised public option in Chapter 7.

The State of South Australia formed the South Australian Water Corporation on 1 July 1995 under the South Australian Water Corporation Act 1994. The Corporation was required to operate to commercial criteria. This included paying interest on loans, and paying taxes and dividends. The South Australia Water Corporation was, and is, a corporatised public utility, and could have operated in a similar way to Melbourne in Victoria. However, the State of South Australia decided that the management of water services in its largest city, Adelaide, should be contracted out. Following competitive bidding, a 15.5 year contract was placed with United Water starting in 1996. This resulted in a rather unusual situation. The South Australia Water Corporation had water service management responsibilities outside Adelaide but not in the main city, and was given responsibility for managing the Adelaide contract. The Corporation also had other responsibilities, including operation of the Murray River on behalf of the Murray Darling River Commission. The organisation structure was as shown in Figure 8.2. The contract with United Water also incorporated a requirement to assist the Corporation in its economic development. This included providing water for exporting outside the state, establishing a research group and funding of significant R&D within the state. The two distinctly different aspects of the contract are identified in the Corporation's organisation chart. I discuss only the operational contract, its initial unpopularity and its performance.

Initially, United Water was owned by Thames Water (47.5%), Veolia Water (47.5%) and Halliburton KBR (5%). In 2005 Thames Water sold its stake to Veolia, so the ownership of the company became 95% French and 5% American. The requirements of the contract were very demanding with United Water required to meet around 180 performance measures [30]. These included measures on drinking water quality, wastewater quality, water pressure and flow, and consumer service conditions. There were penalties for failure to meet these performance requirements. According to the company, these were set higher than the levels being achieved by South Australian Water prior to the contract. United Water was also required to prepare 1 year, 5 year, and 25 year asset management plans, and to submit them annually for approval by South Australian Water Corporation. Decisions on capital expenditure on asset development and refurbishment were made by the Corporation, but with United Water responsible for implementing the programmes. The contract provided for United Water to be paid in three ways. One, there was an annual fixed sum for management and operation of the system. Two, there were reimbursable costs for maintenance and consumer service based on annual targets; savings or overspends were shared. Three, there were payments for design, capital project management and construction management. The contract provided for a five-year price review to take into account any changes in operational requirements.

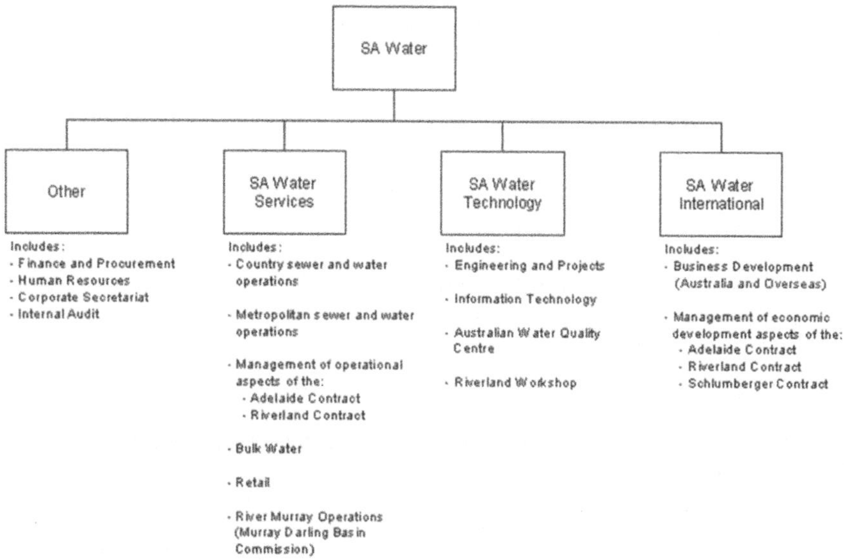

Figure 8.2 Organisation structure of South Australia Water Corporation (*Source*: Government of South Australia).

Just 15 months into the contract there was a three month event, which became known as the "big pong". It took three months for the source of the "pong" to be identified. It was discovered following an investigation, funded by the Government and led by Ken Hartley from the University of Queensland. Hartley found the smell was coming from the largest of Adelaide's four wastewater treatment plants, at Bolivar, 11 miles north of the city. Once identified, the problem was resolved by chemical dosing of the plant's extensive lagoon system. It is reported [31] that Hartley concluded that weather conditions at the time of the pong were no different from previous years, and that the stench resulted from equipment failures and inadequate monitoring. Untreated sewage was being flushed directly into settling lagoons. Taking a neutral stance, it could be argued that the blame should be shared: South Australia Water partly to blame for earlier under-investment, United Water partly to blame for failing to identify the weakness at the treatment plant, and for inadequate monitoring. Publicly, as can be seen from press reports, the blame was placed wholly with United Water as a private company cutting costs and only interested in making profits. The cost of repairs and remediation were shared between the Corporation and the Company, perhaps as internal recognition of failures before and after the commencement of the contract. What will never be known is whether that equipment would have failed, and whether the failure would have gone unnoticed, had South Australian Water Corporation still been operating the plant. Whatever the history, it was a major failure and the responsibility at the time was placed squarely with United Water.

Another bone of contention with the general public, and exploited by those opposed to the use of the private sector, was the reported increase in water tariffs made at the time of the contract. It had been reported widely [32] that the tariff for the first 136 kilolitres (kls) rose immediately (implying on start of contract) by 59%. The actual changes, as obtained from press releases by the South Australian Government, were as given in Table 8.4 below. These figures show that the figure of 59% related to the changes between 1993 and 2001, before the start of the contract. In 2003, the first 136 kls were free but with a fixed service charge of Aus$120. In 1994 the tariff structure changed and the first "free" block was abandoned. I have not given the 136 kls figures for 2006, as they have no relevance in the context of misrepresentation of information.

Table 8.4 Water charges in South Australia 1993–2006, based on tariff figures in Government of South Australia Press Releases.

Year	Cost of first 136 kls			Cost of average consumption of 250 kls		
	Aus $	% increase since 1993	% increase since 1996	Aus $	% increase since 1993	% increase since 1996
1993	120	–	–	220	–	–
1996	172	43	–	275	25	–
2001	191	59	11	301	37	9
2006	–	–	–	335	52	22

The contract began in 1996. The increases in charges prior to the start of the contract were much higher than those afterwards. The decision to change the tariff structure and raise water charges was taken long before the decision to contract out the management of water services. However, there was a significant increase in charges between 1995 and 1996 for the average annual consumption of 250 kls. There was a calculated change from Aus $257 to Aus $275, an increase of 7%. The increase in charges for average consumption over the ten-year period of the contract, from 1996 to 2006, was 22%. I estimate that retail prices rose by 24% over that period, so in real terms there had been a small reduction in water prices. This shows the importance of looking at the facts, rather than relying on loosely reported numbers, which may be highly selective and unqualified.

There are a number of different estimates of the cost savings arising from the contract. According to Grisoti and Saunders [33] the savings were Aus $10 million for every year of the contract. The state Government, as reported in the Adelaide Advertiser, estimated that savings to the Government from the contract would be Aus$66.7 million for the period 1 January 1996 to 30 June 2002. However, those critical of the contract, and perhaps opposed to privatisation in principle, have queried these figures. Some point quite rightly to the difficulty of obtaining reliable information due to the confidentiality of commercial contracts (see *State Secrets* by Spoehr, Quiggin, Wilson and Purse [34]). Stephen Hamnett in his paper *Adelaide as a Sustainable City: Water Management in the Broader Context of Sustainability* [35] says that "validation of (claimed savings) may need to await comprehensive and independent audit of the contract arrangements and achievements at the end of the contract period in 2010". Although there are State Audit General reports on South Australia Water Corporation I am not aware at the time of writing in 2013 of a published audit on the United Water Contract.

United Utility claimed that it met 99% of the 181 required performance standards, and that there were significant improvements in the quality of service. Drinking water quality, which had been poor prior to 1996, with national jokes about Adelaide's tap water, became regarded as being of good quality. The good performance is confirmed in a case study report [36] which gives some key data over the 10 year period from 1996/7 to 2005/6. Some of the key performance figures are also given in the appendices of the Government of Australia's Transparency Statement of Water and Wastewater Prices 2005–6 [37]. Audited figures are published by the Water Services Association of Australia in WSAA Facts [38].

There was a political argument in South Australia about the extent to which the Government of South Australia was receiving dividends on the profits made by the South Australia Water Corporation. In a letter to the Competition Commission in February 2006, Liz Penfold MP requested an investigation into the business practices of the Government of South Australia and SA Water. She referred to the Annual Report of SA Water [39] saying that of the $196 million net profit after tax in 2004/5, there was a dividend payment to the State Government of $155 million (78.8%), and that this would increase to 95% in 2005/6. She related this dividend payment to 39 cents in every dollar taken from consumers in water service charges. This is curious because the annual report shows that costs exceeded revenue and that there was a state subsidy. She also queried the SA Water staff levels, suggesting that the number of 1700 employees appeared to be high, considering that much of the operations work was contracted out. United Water had around 400 employees. She complained about the lack of transparency on SA Water's operations and costs, on the grounds of "Commercial in Confidence". Hopefully, the full facts will emerge at some time in the future.

However, based on the information that is available, the Adelaide contract would seem to have been successful. It delivered improved quality of water, improved service, and cut costs. The "big pong" in 1997 was a major failure by the operator, due to ineffective monitoring resulting in the cause not being identified for 3 months, and then only by an external investigation. Since then there appear to have been significant improvements in the infrastructure, and in operations.

Aside from operational performance, there was the political argument about the payment of dividends to the state, but this was part of the reform process of providing incentives for improved performance, which was being applied right across Australia. There is the question of whether it is appropriate for the Government of South Australia to receive high dividend payments, rather than using the profits for investment in improved services or an improved environment. This is not particularly important, but transparency is vital and this was lacking. If there were to have been good published information, issued by an independent body, giving the terms of the United Water contract, and providing reliable information on the performance of the contractor, there could have been much greater public confidence in the whole system. I believe that public confidence in information requires some form of independent scrutiny, which is a recurring theme throughout this book. It is not possible to have confidence in published information unless it is provided by an organisation which has no vested interest in the results. Neither the body responsible for letting the contract, nor the contractor, meet that requirement.

The United Water contract reached the end of its term in 2011. The company was not included in the shortlist for a new contract [40] which was awarded to a consortium called Allwater. The form of this new contract is very different with emphasis on partnering. Partnering is discussed at the end of this chapter. It involves close co-operation between the two parties and greater transparency. It will be interesting to see how this contract is viewed in the future compared with the United Water one.

Jakarta, Indonesia

In 1996 there were major problems with the water supply in Jakarta. Only 41% of the population were connected to the piped system, and only 30% of those received a 24 hour a day supply [41]. Non-revenue water, largely leakage, was around 55%, and the delivered water was not considered to be potable. In 1997 two concession contracts were let for water services in Jakarta. One was let to a consortium called PT Kekar Thames Airindo, formed by Thames Water, in which a company owned by President Suharto's son had a 20% stake. The other contract was let to a consortium called PT Garuda Dipta Semesta, in which the French company Suez had a 40% stake. At that time, it was considered to be impossible to do business in Indonesia without involving the Suharto family, or their associates in some way. The contracts were signed in June 1997 just before the beginning of the Asian economic crisis. There was escalating inflation and an increasing number of job losses. The situation brought riots to the streets. Following brutal suppression of the rioters in which four student protestors died, there was a national revolt against the Suharto regime, which was overthrown in May 1998. There was controversy over the Thames and Suez executives leaving the country during the riots – the companies said for only one or two days, others said for a week or more. In response to the criticism, the companies said that sufficient management staff had remained to supervise operation of the systems. But according to an investigative report [42], during the period of unrest the operation of the systems was returned to PAM Jaya, the public body responsible for operations prior to the concession contracts.

However, whatever the situation during the period of unrest, the responsibility for operations was resolved afterwards. Both international companies bought out the stakes owned by their previous partners. Thames and Suez formed new companies, PT Thames PAM Jaya and PT PAM Lyonnaise Jaya respectively, which were 95% owned by the international companies, and 5% by their local subcontractors. Due to the Asian economic crisis, there was general turmoil in the country. There was local resistance to the private contracts. As a result, there was a long period of negotiation of new contracts, during which time, although the systems were operated by the companies, investment was delayed. Finally, in October 2001, new contracts called Restated Co-operation Agreements were signed between the companies and PAM Jaya. A regulatory body, the Jakarta Water Supply Regulatory Body (JWSRB) was formed in November 2001. In early 2002 the Customer Community Communication Forum was formed, and in early 2003 five Water Supply Customer Committees were established in the municipalities. This brought in an era of improved transparency, an attempt to inform customers, to facilitate public participation and to assist customers in handling service complaints with the companies. It could have been expected that with the above, effort could have been concentrated on achieving the objectives built into the contract requirements.

What were the contract requirements? They included increasing access from 41% of the population at the beginning of the contracts in 1997 to 100% at the end of the contracts in 2022. There was the expectation that 70% coverage would have been achieved by 2002. Non-revenue water was to be reduced to 25% over the contract period, and the required drinking water quality to be delivered across the whole city by 2010. However, investment commitment figures were not included in the contracts, although investment intentions were built into five-year plans. What was achieved by 2002? According to JWSRB data, in the Thames area in 2002, access had increased to 62%, but in the Suez area it was 41%. PT Thames PAM Jaya reported that by 2003 non-revenue water was down to 42% from the level of 60% inherited, but that the water quality target was unlikely to be met by

2010. By 2003 the level of investment had been only around half of that expected, and even less according to some sources [43] if inflation is taken into account.

Understanding why investment was lower requires knowledge of the process of setting water tariffs and the manner of payments to the water company operators. There is a full and well-researched description of this by Olivia Jensen in her paper to the 4th Conference on Applied Infrastructure Research held in Berlin on 5 October 2005 [44]. In essence the tariffs determined the amount of revenue collected from the consumers, but the two private operators were paid a fixed rate per cubic metre of water delivered, based on the actual costs of delivery. So when tariff increases did not keep up with inflation, the payments to the water company operators exceeded the revenue collected from the consumers. Therefore for some of the time PAM Jaya had insufficient income to pay the private water operators. This shortfall was treated as debt with a growing amount owed to the operators. As a reaction to this, and allowable in the Restated Co-operation Agreement of 2001, if tariffs were not high enough, the private company operators cut back on capital expenditure.

So why didn't tariffs keep pace with inflation and what was the role of the regulator, JWSRB? The regulator did not have the responsibility of setting tariffs; that role was carried out by the Governor of Jakarta. In spite of high inflation, or indeed because of high inflation and the unstable political situation, there were no tariff increases during the period between the fall of the Suharta regime and the Restated Co-operation Agreement of 2001. Also, despite a tacit agreement that tariffs would be adjusted every January, this did not occur on a regular basis. This resulted in uncertainty and in a slowdown in investment.

Privatisation continued to be controversial. There was a resistance to the Water Resources Act of 2004 from various groups organised under the Indonesian Environmental Forum (Walhi), implying that it was a charter for privatisation. It was not described as a privatisation act although a translation of Article 40 (3) reads that "state-owned enterprise and/or regional owned enterprise shall be in charge of the development of drinking water supply system" and of Article 40 (4) reads that "co-operation, state-owned enterprises, regional enterprises, private sector enterprises and the communities may take a role in the development of drinking water systems", suggesting that the Government was keeping its options open. In July 2005, the Jakarta Post reported [45] that a petition by the Indonesian Environmental Forum (Walhi), for a judicial review of the Water Resources Act, had been rejected by the Constitutional Court. Walhi had argued that the legislation would turn water into a tradeable commodity, sold for profit, thus jeopardizing the public's ability to afford tap water. In its verdict, the court interpreted the Act to mean that "the state must take charge of managing water resources, with only limited involvement by the private sector allowed, if the Government fails to do so".

It seemed likely in 2006 that both the international companies, Thames Water and Suez, would pull out of Indonesia. On 28 July 2006 it was reported [46] in the Jakarta Post that Suez had reduced its ownership to 51% by selling shares to its local partner. There was an appeal, by the Chairman of the City Council's Commission on Infrastructure, requesting Thames Water to postpone the sale of 100% of its shares. Thames did pull out at that time but Suez continued for the time being. The difficulty with the antiprivatisation arguments is that they take both political and professional attention away from the important requirement of the delivery of sustainable water services to all. The Jakarta water supply situation in 1996 was very poor; some improvements had been made by 2006, but there was much more to be done to achieve 100% access.

The difficulties with the contracts continued. Suez announced in October 2012 that it would be selling its share to Manila Water, a subsidiary of the Philippines holding company, the Ayala Corporation. This would have to be approved by the city. On 28 March 2013 the Mayor of Jakarta announced [47] that he wanted to cancel the contracts. An audit would be carried out to assess the legal options. A second article [48] appeared in The Jakarta Post on 30 March which expressed concern over possible contract penalties with a suggestion that the city might buy a majority stake in the water companies.

So what are the lessons to be learnt from the Jakarta experience? Firstly based on the early experience in 1996, it might be expected that international companies will wish to avoid unstable political situations, especially where a government is corrupt and contracts require mandatory involvement with the ruling regime. Had there been an open and transparent bidding process, the companies would have been in a much stronger position on regime change, and possibly there would have been less opposition to the use of the private sector. Secondly, although a regulatory body was formed in 2001, it did not have the necessary powers in the tariff setting process, with decisions influenced by political considerations. There should have been a periodic planning process in which forward programmes were agreed, with matching income to pay for them. There needed to be financial penalties on companies for failure to deliver. This required a truly independent regulator. Progress was made on access and level of service (almost 24/7) to the more wealthy areas of Jakarta, but major problems remained elsewhere. The contracts could have specifically targeted the poorer areas, with tariff and other provisions for the poor. The Special Unit for South-South Cooperation produced a case study report on Jakarta [49] in 2012. The report says that "the contracts offer important lessons about structuring contracts for PPPs. Blame for the failure of the PPPs cannot be placed solely on either the private or public sector; both sides had a role in the limited success seen. Choosing the right partners and creating a good working relationship is almost as essential as creating a clear and straightforward contract."

It is not clear at the time of writing in 2013 what will happen next. Will there be an international presence perhaps from The Philippines rather than France, or will there be ongoing contracts with local private firms, or will there be something similar to a corporatised public operation? Whichever way it goes, the basic problems will be the same. Delivering 24/7 for all will require clear policies, sound planning, investment in both the existing and new infrastructure, public involvement and transparency, and sustainable cost recovery. As can be seen from the situation in Bolivia discussed earlier in the Chapter, simply removing the private sector won't in itself address any of these requirements.

PRINCIPLES OF EFFECTIVE PROCUREMENT

The problems experienced above can be attributed to a combination of lack of effective regulation and lack of effective procurement. In principle the approach to choosing a private sector company, to manage the operation of a water service, should be no different from selecting a contractor to build something, or an IT company to provide computers and information systems. However, there have been a number of highly publicised problems in the UK, such as the construction of Wembley Stadium, and provision of software for the Health Service, to illustrate that problems in procurement are not limited to the rather more emotive area of water service operations. Before considering sound procurement principles I list a number of different types of contract in the area of water services. Afterwards I focus my attention on the considerations necessary for successful concession contracts. The types of contract can be categorised as given in Table 8.5.

Generally the principles for successful procurement apply equally to all types of contract. As effective procurement is a large subject, it can only be summarised in this book. For wider reading see [50] and [51]. A vital first step is for governments and utility managers to recognise that procurement and contract management are different from local government and direct operations management. Those without experience of procurement should seek expert advice. There is a need to avoid any pre-conceived views that private companies are by nature aliens, crooks or enemies. Yes, companies are in business to make money, but successful companies do that by providing good value to their clients in an honest way, without which they will not remain in business for long. Contractors should be encouraged to share the client's values and objectives, so that there are incentives for win-win situations with sharing of benefits. The initial mindset is important in establishing a business climate of partnership, co-operation and openness. Much of that required for successful procurement is common sense. However, as there are so many failures, it has to be assumed that frequently the fundamental requirements for success are lacking.

In preparing a specification for contract bidding, there should be clarity of objectives, and clarity on what is included and what is excluded. Generally, outputs should be specified so that contractors can provide the benefit of their knowledge and experience. In the case of water treatment plant, specifying particular processes limits the scope of getting the best out of contractors. For this reason, I prefer DBO (Design, Build, Operate) to BOT (Build, Operate and

Transfer) contracts. I prefer full concession contracts to either of these more narrowly-focussed contracts, and I return to that later.

Table 8.5 Types of contract in the construction and management of water services.

	Functions covered
Types of contract for whole systems	
Management	This involves only the operation of an existing system without any capital investment in either replacing existing assets or providing new ones
Affermage	This French term is used for a combination of operating and investment in the refurbishment of existing assets
Concession	This term is often used loosely to include also the two types of contract described above. A true concession contract involves the operation and refurbishment of existing assets and investment and operation of new assets
Types of contract for part systems	
DB (Design and Build)	This is limited to designing and building a new plant, or part of a plant, which on completion is handed over to others to operate
DBO (Design, Build and Operate)	Added to the above is the responsibility of operating the plant for a specified period after which it is handed over to the final owner, usually the public authority
BOT (Build, Operate and Transfer)	This is as a DBO except that the plant has been designed by others
Multi-systems	
Partnering	This can be applied to a single construction but generally involves a large programme of construction as in Scotland in the UK. It can also include both construction and operations as in the highly successful case of Wales in the UK. Both are described later in the chapter.

It is important that sufficient time is taken to assemble the best information available, and to identify areas of significant uncertainty. Where there is uncertainty on the condition of assets, such as on the condition of the Atlanta distribution system described earlier, there is a need to build into the contract effort to obtain the necessary information, and to make provision for the contract terms to be adjusted accordingly at a later date. It is highly desirable that a draft specification is published. It provides transparency at the outset. It should also be presented in such a way that potential contractors are able to obtain clarification on requirements, and also to offer suggestions on improvement of the contract conditions. There should be sufficient time for this aspect, to ensure that there is clarity on the requirements of the contract, and that there is no "fudging" of the issues. There can be a temptation to "fudge", (that is to gloss over difficulties), in an attempt perhaps to meet some political deadline, but it is certain that eventually it will be necessary to address seriously overlooked problems. Inevitably, such contract deficiencies lead to higher costs, and have the potential for disputes between the parties and the potential for public criticism.

This raises the question of risk. Appropriate allocation of risk between the parties is key to success. There can be a tendency to require the contractor to take all the risks, but that approach does not make the risks go away. Either the contractors will include the accommodation of risk in the prices, or if a contractor is unable to live with the consequences, a contract will fail and the public authority will have to cope with those consequences. It is important that risks are allocated and borne by whichever party is best able to manage them.

Risk is associated with uncertainty and change. Contracts are generally for ten years or more, and it is important to recognise that circumstances will change over that period. Contracts need to be adaptable, with built-in procedures for addressing changing circumstances. It is better to have a partnership approach to change, with the parties working together to find the most cost-effective solutions. Contract "scoring" in which either party tries to win to the detriment of the other should be avoided, as conflicts which result in legal action only benefit lawyers. Should there be a major change in circumstances, there may be a need for immediate adjustment to the contract conditions. However, a number of small changes can be dealt with in a Periodic Review, perhaps every five years, when both requirements and contract payments can be adjusted in an integrated way. Such reviews are best overseen by an independent regulator, but more on that later. If there isn't an independent regulator, who can act as an arbitrator? Some other form of arbitration procedure should be agreed by both parties, and included in the contract. Should it be necessary to terminate the contract, perhaps due to a change in the politics of a country, or perhaps due to failure by the contractor to meet measurable outputs, there should be a termination procedure which is fair to both parties. In this respect, both compensation and penalty clauses may be included in the contract termination procedure. There are essentially three types of changing circumstances, which I consider in turn:

- Political/governmental;
- Legislative and currency changes;
- Physical aspects.

An example in the political/governmental category is a total change of attitude to the use of the private sector for water service operations. In Indonesia, some of the difficulties arose from the anti-Suharta backlash when the regime fell. In Bolivia, the change to a socialist Government has resulted in the termination of all concessions, whether or not they were working well. The same thing could have happened in Chile, but whilst the incoming socialist Government stopped the sale of the operating infrastructure to the private sector, it decided to continue with the use of the private sector through a concessionary approach. A reversal of policy can result from a change of mayor, as in Atlanta. Even changes of minister within the same government can trigger changes, as can reorganisation of government departments with a change of senior officials. The emergence of new anti or pro something bodies, most commonly anti-private pressure groups, can have a major impact on public opinion and thus government policies. These risks need to be assessed, particularly in countries with a history of political instability.

Changes in legislation can and do occur over the period of a contract. The most common ones are new drinking water quality standards, or new environmental standards, requiring increased investment. Generally, these can be predicted within a five-year cycle and provisions for the changes can be handled during the periodic review process. Handling such changes, which normally require an increase in water tariffs to pay for the additional investment, requires good communication with the general public, to ensure that they understand the reasons for the increases. If there is a communication vacuum, such changes provide the opportunity for anti-private groups to blame private sector "profits" for increases in tariffs made by governments. Other examples of legislation changes are changes to tax regimes and changes to general legislation impacting on procurement rules. Currency changes can be critical, especially if the cost of operation is influenced by another currency, such as the import of vital components or consumables. Where possible, every effort should be made to source locally.

Making provision for adjustments for physical changes is very important. The Atlanta case demonstrates how important it is to understand the condition of the systems, so that operating costs can be estimated with some accuracy. Where there is a lack of knowledge at the time of letting a contract, provision should be made to obtain the required information, and to adjust the contract conditions accordingly. A very common area of uncertainty is water distribution system leakage. It should be possible to know the level of non-revenue water fairly accurately, as it is the difference between the amount of water going into supply and the amount of water for which revenue is collected. However, in systems which have not been properly maintained, there are no reliable measurements of the amount of water leaving treatment works. This relies on having accurate meters at treatment works. This can be assessed fairly easily using insertion meters, but the proportion of non-revenue water, which is leakage, is more difficult to determine. It is vital that a full water balance [52] is carried out either before the contract is written or, failing that, to make it a condition of contract that it is carried out early in the contract. Reducing the level of non-revenue water is often one of the key objectives and performance measures, so having an accurate understanding of where all the water is going is essential. Another possible change in physical conditions is the quality of source waters, such as quality changes due to increased pollution. The provision to make contract adjustments for changes that are outside the control of the contractor should be included. Equally, should source water quality improve to the extent that the treatment costs reduce significantly, the reduced operational costs should be taken into account at a periodic review.

I have mentioned that the allocation of a risk in a contract should be based on which party is best able to manage it. The contractor should bear the risks associated with operations, and financial risks associated with construction projects, all of which should be under its control. Assuming that the contractor can operate without political interference, and has the ability to set up communications with the media and consumer groups, the contractor should be allocated risks associated with customer relations. This gives an incentive to provide a good service. If it is government that is letting the contract, whether national or local, it should undertake those risks associated with the impact of new regulations or tax changes, risks related to deterioration in source water quality, and risks associated with significant currency exchange rate changes. It is important that all these aspects are included explicitly in the contract to avoid subsequent arguments.

Establishing well-defined and well-drafted contracts requires knowledge, experience and sufficient time to work through all the issues. Extra time spent at the preparation stage can make the difference between success and failure, and avoid time, effort and cost later in dealing with contract deficiencies. It is important that, before inviting bids, the authority letting the contract should have a good estimate of the likely range of contract costs, and have well-defined selection procedures that take into account both cost and quality. The final contract specification should be checked carefully to ensure that all the issues have been addressed fully, and that no aspects have been "fudged". The need to avoid glossing over difficulties cannot be stressed too much.

PARTNERING

In recent years partnering has taken much greater prominence in the UK through the need to achieve very large construction programmes on time and to budget. Partnering is just that, partnering, in which two parties enter into an agreement to deliver something in an administratively efficient way, within a climate of trust, to avoid the conflicts that can arise from more traditional contracting, when the only beneficiaries are lawyers.

In partnering there are no invitations to tender for specific contracts. There are no conventional construction contracts. Instead, the customer and chosen suppliers enter into a partnering agreement, in which there is a contract framework with target prices. There is a so-called pain/gain approach through an open-book arrangement. If the contract outperforms, the partners share the gains. If there are problems and additional costs, the partners share the pain. Therefore there is a joint incentive to succeed. It relies on each partner understanding the other's goals, and sharing a commitment to achieve each other's goals to establish a win-win situation. It involves total transparency, sharing information and solving problems together. Instead of looking to blame each other, there is a culture of "how can we solve this problem together?" There has to be complete trust by each partner, with a demonstration of the required open behaviour to engender confidence in each other and in the partnering process. The approach is to start with facilitated workshops to establish the goals and the plans to achieve them, so the supplier is part of developing solutions right from the beginning. When problems arise, there are further workshops to find solutions to those problems. The management processes relate to managing the framework agreements, rather than monitoring the minutiae of individual contracts. The largest partnering agreement of which I am aware is for the delivery of major water and wastewater construction programmes in Scotland. There had been decades of under-investment in water services in Scotland, resulting in the following circumstances in 2002:

- 78 wastewater treatment plants failed compliance with discharge standards;
- 40% of wastewater plants were considered "unfit for purpose";
- 60% of water mains and sewers were judged to be in poor or very poor condition;
- There were 9600 burst water mains every year.

Scottish Water, the corporatised-public water services company covering water and wastewater for the whole of the population of 5 million people in Scotland, estimated that £2.3 billion needed to be invested during the period 2002–6. The economic regulator, at that time the Water Commissioner, said that the Company did not need that amount and that it could achieve the required programme for £1.8 billion. As an indication of the scale of this lower figure, it amounted to around £360 for every person in Scotland. Scottish Water decided that the very large programme could not be delivered by conventional contracting, due to the sheer magnitude of letting such a large number of individual contracts in a short time period, and the need to complete the work at a total cost of 22% less than the estimated figure. The company therefore decided to use a partnering approach and formed Scottish Water Solutions. Scottish Water Solutions is 51% owned by Scottish Water, with consortia led by Thames Water and United Utilities owning 24.5% each. The consortia are made up of 8 companies, including large UK contractors. The Partnership was required to deliver 2300 projects in 2.5 years, amounting to 40% of all construction in Scotland during that period. By March 2004:

- 813 km of water main had been laid;
- 103 km of sewer had been replaced;
- Construction had been completed at 59 water treatment works;
- Construction had been completed at 20 wastewater treatment works;
- An additional inherited programme was 80% complete ahead of time and at lower cost;
- There were significant savings on the new construction programme.

One year later, the figures given in the 2004–5 Annual Report [53] were as follows:

- A £364 million investment programme had been delivered in that year;
- The regulatory output targets had been exceeded by 4%;
- 1176 km of water main had either been refurbished or replaced, setting a new UK record;
- Leakage had been reduced by 44 megalitres/day, equivalent to 6% of production;
- 156 km of sewers had been refurbished or replaced;
- 79 wastewater treatment plants had been either built or refurbished; and
- There had been improvements in drinking water quality and improved pressure in parts of the distribution systems.

In 2005–6, a £468 million investment programme was delivered, an increase of 28% over 2004–5, with the following impressive statistics:

- Rehabilitation of 2774 km of water main;
- Rehabilitation of 337 km of sewers;
- Solutions to 322 unsatisfactory combined sewer overflows.

Scottish Water Improvements Report [54] gives information on what was delivered over the period 2006–2010. The capital programme which amounted to £260 per property per year over that period had peaked at £700 million in the year 2008/9. The £1.8 billion programme had largely been delivered on time. Some of the larger projects could not be completed in the time and some had been subject to planning delays. Most of the projects were to have beeen completed by the end of 2010. It is necessary to be a little cautious about that original figure of £2.3 billion, as it could have been part of the "regulatory game" of over-estimation. However, it is certain that the Regulator's allowance of £1.8 billion would have been very demanding. However, the report said that with experience of the capital programmes since 2002, investment should not exceed

£500 million in any one year and that this would be taken into account in future programmes.

Scottish Water Solutions embarked on another major programme of partnering, this time with a different consortium and involving only above-ground asset construction. Water main and sewerage (underground) work would be handled internally. Although smaller than Scottish Water Solutions I, partly because the underground work was excluded, the partnering contract [55] involved 280 projects costing £450 million over the period 2010–2015.

I include this example, not simply to give an illustration of successful partnering on a large scale, but also to show how the public and private sectors can work together in a noncontroversial way. Scottish Water Solutions is an example of an effective public-private partnership.

A PERFECT CONCESSION CONTRACT?

Utopia in outsourcing water service operations and development might be considered to be the "perfect" concession contract. Can such a perfect contract exist? Perhaps not, but it is possible to let very good contracts which have an excellent chance of avoiding the problems experienced in the examples discussed in this chapter. I see the following key ingredients:

- A concession of a size large enough to attract good management and to give economies of scale, which probably means a minimum of 250,000 connections. This need not be within one hydraulic system, but could be the management and operation of a number of systems;
- A contract period of around 15 years. This is long enough to achieve management commitment and some continuity, but short enough to provide for the potential benefit of re-tendering. Contracts involving a high level of investment by the contractor may need to be longer, to alleviate the impact on charges by providing for cost recovery over a long period of time;
- Five-year periodic reviews;
- Clarity of requirements with clear objectives, sound information on condition of assets and reliable revenue forecasts;
- Good consumer consultation beforehand with a requirement for the contractor to handle consumer relations;
- The contractor given full responsibility to manage with no outside interference;
- An independent regulator to oversee the letting of the contract and to oversee the periodic reviews. Failing this, the naming of an agreed arbitrator in the contract;
- Tariffs based on need only, and not increased to fund other local services; and, above all;

- Transparency, with contract conditions published with good information provided to consumers. This is particularly important if the introduction of a contract coincides with increased tariffs in a move to sustainable cost recovery.

REGULATION IN CONCESSION CONTRACT SITUATIONS

In Chapter 2, I discussed "regulation by contract" saying that a well-defined contract provides the means of controlling the delivery of the desired performance of a water service operator, whether public or private. I said that it is a major step forward, and much better than the "traditional" municipal operation in which there are limited or no performance measures, and poor financial ring fencing. But I stressed that it does not provide any means of regulating the contract owner, nor does it provide for transparency and public participation. I went on to say that, on the contrary, it tends to reduce transparency, as it assumes that governments can look after the consumers' interests. This lack of transparency risks the creation of the conditions for corruption. So some form of independent oversight is required. This could involve regulating the contract owner. Dŵr Cymru is the "public" licensed "water undertaker" providing water services to Wales, and is regulated in the same way as the private companies in England. During the period when its operations were contracted out it continued to be the regulated utility. It had to take full responsibility for any failures by its contractors. Although the Dŵr Cymru situation is rather unique and the economic regulator was not involved in the contracts, the deliverables were defined by the regulatory system (See periodic planning in Chapter 3). More generally, without such a rigorous regulated planning system, an independent regulator can ensure that the conditions, on which a contract is to be based, are clear, objective and realistic. I believe that effective independent regulation could have avoided many of the problems described in the examples given earlier in this chapter. I do not see regulators interfering with the contracts, but providing the objectivity and transparency on which the contracts are based.

EQUITY-BASED PRIVATISATION

Attention so far has been focussed on the use of contractors to provide services in one of a number of forms. In none of the cases discussed does the ownership of the operating assets rest with the contractors, as even with concession contracts new assets are transferred to the local public body at the end of the contract. Equity-based privatisation involves the sale of assets to private companies. These companies are licensed to provide water services for specified periods of time. The licence includes the requirement to maintain the serviceability of existing assets and to finance, build and operate any new assets that are required to meet defined objectives. The story of England and Wales is given in Chapter 9. Some other examples of part-equity based privatisation are described below: that of Chile, within a strongly regulated environment, and some examples in China, as part of its marketisation programme.

Chile

Chile (see Figure 8.3) is an important example because, through strong regulation and good social provisions, Chile is one of the most successful countries in Latin America in the provision of water services. It is a story of a number of stages of reform, the last stage being the involvement of the private sector to bring in much-needed investment in the wastewater infrastructure.

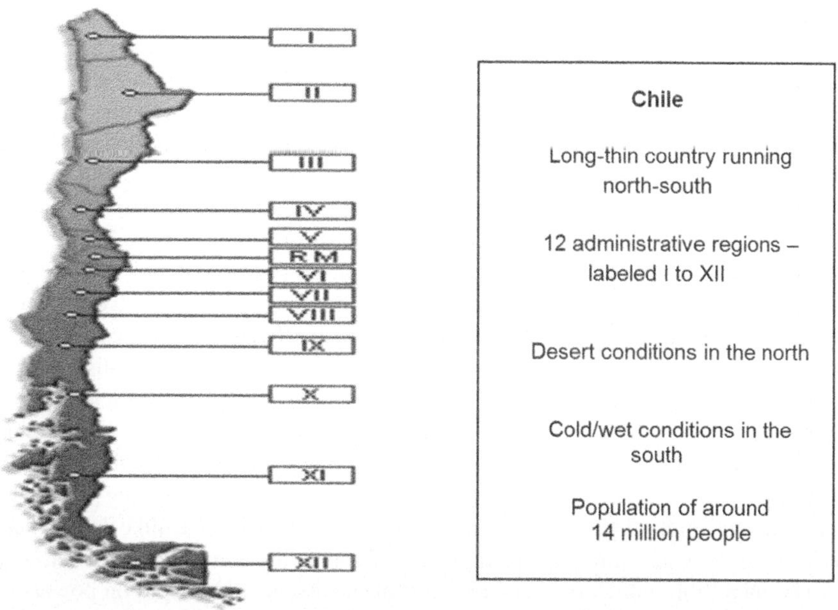

Figure 8.3 Map of Chile and some aspects of climatic conditions in the country (*Source*: Adapted from map presented at AIDIS Conference held in Joinville, Brazil 2003).

The story begins in the 1970s when the situation was very different from today. During the Allende period of 1970–3, when it was considered by the Communist Government that essential services should be free, water charges recovered only 5% of the cost of water supply. There were no incentives for consumers to use water wisely, there was insufficient money to maintain the system, and the condition of the infrastructure deteriorated. Until 1997 water services were organised by the regions, but following the change in the political regime, the Government decided to bring all water service operations under one national company. Later there was a move back to individual companies in each of the administrative regions, but with all of them coming under the overall management of a holding company, Corfu. A policy of gradually increasing charges, towards cost recovery of operations and maintenance costs, was introduced.

In 1998, there was the Tariff Law, which set-up an economic regulator Superentendencia de Services Sanitarios (SISS), and reformed the approach to setting water charges. The new tariff framework established four basic principles: those of economic efficiency, full-cost recovery, non-discrimination between consumers, and clarity on managing demand and consumption to encourage conservation. SISS had little autonomy, being dependent on the Ministry of Finance for funding, and no real powers, having to rely on Corfu for enforcement. However, it made some advances in regulation by creating a "model" company against which the thirteen companies (one from each region, plus the company serving the capital, Santiago), could be compared. This model calculated the efficiency improvements which could be expected, and SISS took efficiency measures into account when calculating tariffs.

Under this regulatory regime, some of the companies made significant improvements in operations and in financial performance. Leakage levels were reduced and some companies achieved full recovery of operations and maintenance costs. But improvements were not universal, with Corfu being unsuccessful in dealing with the management of inefficient companies. With the need for further improvements, the Government decided to establish strong independent regulation, and then to involve the private sector. SISS was given enhanced powers, and became a wholly independent regulator. Its funding became based on a levy on water companies, similar to the economic regulator, Ofwat, in the UK. Government saw the private sector as bringing improved management, and necessary investment in the infrastructure. Unlike in England and Wales, where all the equity was floated on the stock market, the Government, through Corfu, retained 35% of the assets, with a veto on powers of transferring assets. Later, with a change of Government, the form of privatisation was changed from equity sale to concessions, so that ownership remained wholly public. The combination of equity sale and concession contracts raised $1 billion, which was invested in the infrastructure. A significant amount was invested in sewage treatment, a requirement from the European Union as part of the conditions of a trading agreement.

There are some similarities between SISS functions and those of Ofwat in the UK. They have similar powers, including those to set tariffs. They both monitor development plans and service standards. They resolve disputes between companies and consumers. However, SISS does a lot more, including monitoring concession contracts, establishing and monitoring compliance with technical standards, and having responsibility for the control of discharges to the water environment. SISS does not rely wholly on output measures, but also uses input controls, such as certification of water company operatives, in order to raise standards. Perhaps therefore, it is a good model for consideration by countries which are at an earlier stage of development than Western Europe.

With a move to full-cost recovery there are, quite rightly, concerns over the poor. The concern is usually over the ability to pay, not recognising that the poor are often paying far more to water vendors than they would if they had a piped supply. There should be concern, but this should relate to the lack of ability to save, and thus the inability to settle accounts in single payments, such as with monthly or quarterly bills. Chile has gone some way to recognise the need to "pay as you go" through easy payments for the cost of access. Chile adopted a very progressive social policy in relation to water services with priority given to providing access to the poor. Between 1987 and 1995, half of all new connections were for the poor, with local authorities funding the necessary extensions to the distribution systems, and new consumers being able to pay for the connection charge in monthly instalments over 5 years. This consideration was based on the amount of money that these poor people had previously been paying to water vendors. In the achievement of full-cost recovery from charges, general subsidies were converted into direct subsidies to the poor to assist in their payment of water bills. The water companies bill all consumers the full amount but those who, from a means test system are regarded as eligible for assistance, can take their bills to their local authority. The authority pays part of the bill providing the consumer pays for their part. This system has a number of important features. Firstly, the water company can operate in a wholly commercial way, not having also to serve as a social services department. Secondly, consumers see the full bill for the water they have used, so are aware of the full cost of the service they have received. Thirdly, the consumer's contribution to the payment is usage dependent. The subsidy system works well but there is a cost and administrative burden in maintaining a means-tested register of consumers.

Some interesting studies have been carried out on the transformation of the water sector in Chile. The paper [56] by Bitran and Valenzuela, comparing private and public performance, gives some interesting statistics on capital expenditure, charge rates, efficiency and unaccounted-for water. The paper refers to the noticeably higher level of investment by the private sector. The reason is considered to be smaller public sector operations, and the inability of those utilities to persuade Government to make capital available. It is concluded that the remaining state-controlled companies may be privatised in the longer term. However, the authors warn that although in the longer term the more efficient private companies are able to reduce charges, there are necessary short-term increases which could lead to a significant obstacle to privatisation. A 2004 United Nations Research Institute for Social Development (UNRISD) study ([57]

English draft and [58] Spanish final paper) concluded that privatisation of water and sanitation services has been possible due to:

- The existence of consolidated infrastructure systems and adequate laws in the country;
- Stable political conditions ; and
- A well-structured and reliable regulatory system.

However, the study is sceptical about the need for privatization, referring to increases in charges greater than the cost of living and suggesting that, if the state had retained ownership, the increases would have been smaller but still sufficient to achieve full-cost recovery.

The paper [59] by Shirley, Xu and Zuluaga gives an interesting account as to why the EMOS, the water service provider serving the capital, Santiago, was not privatised initially. EMOS, under the same regulatory system, operated under a contract similar to that used in the private concessions. It was later privatised but the paper suggests that major improvements in performance had been achieved as a public sector operation, and that privatisation may not contribute to significant additional improvements. As described above, there were means-tested subsidies to assist the poor. The paper concludes that such (concession type) contracts provide all the elements to establish commitments, and give incentives to managements to make improvements. The paper warns against viewing Chile as a model for developing countries, saying that success in that country has been in part due to its constitution. The constitution supports a contractual approach with targeted subsidies. Its form of public sector governance prevents corruption, and a long history of private water rights has provided a public acceptance of water as a tradeable private good. However, it is important that this warning does not detract from Chile being seen as a good model. Chile has in place all the important elements for success. It has strong regulation providing transparency and driving efficiency. It achieves full-cost recovery through providing targeted subsidies to look after the poor. As subsidies are handled by the local authorities, the companies have been able to concentrate on improving services and efficiency without having to act as a social service. This has resulted in an improved service to the poor.

In 2010 there was an earthquake in Chile with damage to the infrastructure, which might have been expected to have caused major cost problems. However it was reported [60] that an additional US$100 million investment in the infrastructure was required over the period 2011–2015 (an increase of 20% capex) but that it would have a neglible impact on tariffs. This is partly because the concessionaires were insured against earthquake damage. Nevertheless it is a good indication of the robustness of the water services industry in Chile.

Each year SISS produces a comprehensive management report. An example of the information provided in its 2011report [61] translated from the Spanish is shown in Table 8.6.

Table 8.6 Performance Data Chile.

	2011	2010	Variation
% of interruption of the water supply service notified to the customers	34.4	32.4	6% increase
Number of customers affected by the interruptions of the drinking water supply	414	161	157% increase
Average duration of interruption to water supply (hours)	3.5	5.4	35% reduction
% of interruptions to wastewater service as a result of actions by the service providers	77.4	85.7	8% improvement
Average number of customers affected by the interruption of the wastewater service	7	5.6	25% increase
Average duration of the interruption of the wastewater service (hours)	6.7	7.6	12% reduction
% compliance with drinking water quality regulations	98.1	99.4	1.3% lower
% compliance with waste water discharge regulations	97.5	99.1	1.6% lower
% errors in water service bills	0.5	0.9	44% improvement
% of customer claims accepted by the service providers	66	71	7.0% reduction

Source: SISS Management Report 2011.

China

China's phenomenal economic growth is well known and has been studied extensively. Less well known is China's reform of its public sector through a process called marketisation. This involves loosening the grip of Government, through establishing a form of corporatised public company, and through use of the private sector. Private sector involvement has been through a range of different forms of contract. This has included BOT and full concessions. The private sector has also taken an equity interest through joint ventures with cities. There have been both successes and problems. I describe briefly a successful concession contract, and a BOT contract that has brought major financial problems to a city. I then describe a successful joint venture coupled with a concession contract which is showing good early results. Tsinghua University studied 17 private sector cases which in total cover water and wastewater services to around 50 million people. The results are published in a book [62] which provides useful lessons to be learnt on privatisation in China.

Macao

The first Macao concession contract (for 25 years) began in 1985, before Macao was returned to China. At the end of the first contract it had delivered a good quality water supply service to Macao, with the following results [63]:

- Drinking water quality reached EU directive level;
- During the period 1982–1998, water consumption tripled, but this was matched by increased supply capacity;
- Investments over MOP 666 million (around US$ 83 million) and a 20 percent reduction in water charges were achieved;
- Eighty-five percent of distribution system was replaced;
- The level of non-revenue water was reduced from 40% to 20% during the first three years of the concession;
- The leakage level was 12.4 percent in 2001;
- Through increased productivity, the number of employees remained at a total of 360 even though water revenue increased threefold;
- The utility operation became ISO 9001 (Quality Management Standard) accredited;
- Customer service was greatly enhanced through the establishment of a one-stop shop communication facility, and through the formation and operation of a consumer group.

Table 8.7 Sample of Macao Water Performance Data.

Item	Unit	2010	2011	Variance
Installed Design Capacity per Day	'000 m^3	330	330	0.0%
Peak Day Demand	'000 m^3	239	236	−1.4%
Total Pumpage	'Mm^3	77	78	1.1%
Non-Revenue Water	%	12.5	8.8	−29.9%
Volume of Imported Raw Water	'Mm^3	76	79	4.2%
Metered Consumption	'Mm^3	67	71	5.1%
Length of Network	Km	595	600	0.8%
Annual Consumption per Capita	m^3	122	127	4.1%
Annual Consumption per Domestic Accounts	m^3	179	183	2.2%
Total Accounts	'000	205	210	2.7%
Total Headcount	Nos.	257	258	0.4%
Capital Investment	10^6MOP	50	53	4.4%

Source: Annual Report 2011.

All the above is good news. However, I dislike the form of contract, which was a cost plus management contract. This type of contract gives no incentive, other than reputation, for the contractor to strive for improved performance, nor does it provide a means of sharing any cost saving benefits between the city and the contractor. I would prefer to see a contract with annual payments based on need and performance, with some independent scrutiny of performance through comparison with other cities. However, in this case it appears to have worked well. It has not courted adverse publicity and has not been subject to much academic interest. It has produced good results and was renewed in 2009 to run for a further 20 years. Macao Water publishes annual reports. An example of the performance data in the 2011 report [64] is given in Table 8.7. There appears to be renewed emphasis on reducing leakage.

Chengdu

Chengdu, a city with a population of 2.46 million people, is in Sichuan Province and is located around the upper reaches of the Yangtze River. In 1999, Chengdu Municipal Government entered into an 18-year BOT contract for the construction and operation of a new water treatment plant. This involved a 2.5-year construction phase and a 25.5-year operation period. The capital cost of the plant was $107 million with the contractors providing 30%, and the remainder coming from the Asian Development Bank, the European Investment Bank and five commercial banks. The arrangement is that the contractors receive a fixed amount of income each year for the delivery of an amount of water based on demand estimates made in 1999. The contract does not include any provision for review and, at the contract planning stage, no consideration was given to the operational need to balance the input of water from the various treatment works. This would have allowed the city to use water flexibly to achieve optimum operational and cost performance. As a consequence, with demand below the 1999 expectation, the city is paying for more water than it needs. Chengdu Water Company, which has a reputation for sound management, has changed from being a profitable enterprise, to one that is making a loss. The risk to the contractors was low, especially as they provided only 30% of the finance, with most of the risks falling on the city. I am prejudiced against BOT contracts, as they are necessarily sub-optimal. It is far better for the whole operation of a city to be integrated to achieve the most efficient use of water resources. The Chengdu example is one to be studied by other cities, and avoided.

Ma'anshan

Ma'anshan is a city with a population of 700,000 people located in Anhui Province in the east of China. This example is one of staged involvement by a Chinese private-sector company. It entered into partnership with the city, and has provided both investment and management expertise. It states that it has achieved an enterprise culture with the prospect of creating a world-class utility. It began in 2002 when Beijing Capital Company and the city of Ma'anshan water company, Ma'anshan Water Company, formed Ma'anshan Beijing Capital Co Ltd on a 60%/40% basis respectively, to build additional water treatment capacity. This proved to be successful, and in 2004 the joint venture was extended to cover the whole system. The structure of the joint venture is shown in Figure 8.4. Beijing Capital Company is an investor, and provides management expertise. The city, through its water company, maintains ownership of all the assets. The distribution network is leased by the city to Ma'anshan Beijing Capital Co Ltd. There is regulation through the local Construction Committee, which considers tariff applications. The company operations were restructured and, as part of improving performance, key performance indicators for each department were introduced. This was accompanied by training for staff employed on contract through a competition process.

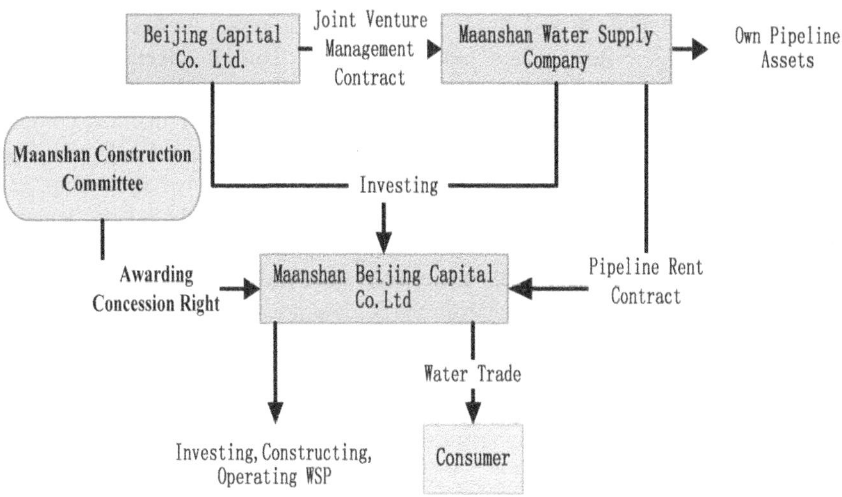

Figure 8.4 The Ma'anshan joint venture structure (*Source*: Tsinghua University, China).

The achievements (personal communication) of the joint venture are listed below:

- The company worked with the city in developing a forward plan for the period 2008–2020.
- Total assets grew in value from 225.7 million yuan (around US$ 37 million) in 2004 to 623.41 million yuan (around US$ 101 million) in 2012.
- The national standard for drinking water quality was achieved. This had required upgrading the technology of the original plant and building a new treatment works.

- From 2004 to 2012, the total water supply increased from 78.3 million cubic meters to 109 million cubic meters. In 2012 there were 260,000 connections.
- By 2012 all customers were metered with smart (remote) meters for 150 large customers. Revenue collection increased from 70% to 99.3%.
- GIS (geographic information system) and GPS (global positioning system) technology are used in distribution network management. District metering (DMA management) is used as part of leakage control management.
- By upgrading pumps, electricity consumption reduced from 268.55 kwh/km^3 in 2004 to 238.01 kwh/km^3 2012.
- Introduced modern SCADA system to manage plant and networks, and introduced pipeline network modelling techniques to support system planning and design. The response to this point might be "so what", but so often poor information on underground systems prevents effective analysis, and severely limits the ability to manage operations and to plan for system expansion.
- Established a 24-hour customer service call centre as part of improving customer service.

In Anhui Province, water services are ranked based on water production, water sales, service access coverage, sales income and drinking water quality. In 2012 Ma'anshan was 3rd among the 22 municipal water companies in the province. The Ma'anshan-form of joint venture had become regarded in China as a successful approach to the reform of water services, based on the Chinese term of marketization.

THIS CHAPTER

In this chapter I have attempted to highlight the important lessons on the use of the private sector in each section. In chapter 10, I bring the various aspects from this and the other chapters together to give general conclusions.

REFERENCES

1. Trench R. and Hillman E. (1984). London under London. John Murray, London, ISBN: 9780719552885.
2. Flaxman T. and Jackson T. (2004). Sweet & Wholesome Water. Flaxman & Jackson, England, ISBN: 9780954898601.
3. Payen G. (2004). Local Management and Competition. IWA Workshop, Tokyo.
4. Egghardt H., Lammerhuber L. and Riha G. (2003). The Water of Vienna. Bohmann. ISBN: 9783901983221.
5. San Antonio Water System History and Chronology. www.saws.org (accessed July 2013)

6. Bayliss K. and Hall D. (2002). Unsustainable conditions – the World Bank, privatisation, water and energy. August 2002. www.psiru.org (accessed July 2013)
7. Bank 'soul searching' on privatization. April 2003. www.brettonwoodsproject.org (accessed July 2013)
8. United Water and City of Atlanta Sign Nation's Largest Public-Private Partnership for Water Operations. City of Atlanta Press Release November 1998.
9. Water Privatization Becomes a Signature Issue in Atlanta, in the Atlanta Journal and Constitution. February 2003, www.ajc.com (accessed July 2013)
10. Segal G. F. (2003). Pulling the Plug on Water Deal a Mistake, in Atlanta Journal and Constitution January 2003, www.ajc.com (accessed July 2013)
11. Operations & Performance Report. Department of Watershed Management, City of Atlanta, March 2006.
12. Performance Audit 2009. City Audit Office, City of Atlanta, www.atlantawatershed.org/audit/e2.PDF (accessed July 2013)
13. Performance Audit Response. Department of Watershed Management, City of Atlanta, 2009.
14. Review of Strategic Financial Planning Model: City Audit Office, City of Atlanta, September 2012.
15. Buried No Longer: Confronting America's Water Infrastructure Challenge (2012). American Water Works Association. http://www.awwa.org (accessed July 2013)
16. Nickson A. (2002). Organisation Structure and Performance in Urban Water Supply: The Case of the Saguapac Co-operative in Santa Cruz, Bolivia. School of Public Policy, the University of Birmingham, Birmingham, UK.
17. Bolivian Government Signals Contract End, in Water 21, 25 January 2005.
18. Chang J. (2006). *Bolivia's Water wars Coming to End Under Morales. The Mercury News*, 26 February 2006. www.mercurynews.com (accessed July 2013)
19. Cuba Jorge. Free or foreign: the water battle in Bolivia. March 2002. http://www.mindfully.org/Water/Bolivia-WaterPrivatization16mar02.htm (accessed July 2013)
20. Bechtel Perspective on the Aguas del Tunari Water Concession in Cochabamba, Bolivia. Bechtel Press Release, March 2005. www.bechtel.com/newsarticles (accessed July 2013)
21. 'Private passions'. The Economist. 17 July 2003.
22. Chávez Franz. Bolivia: Cochabamba Still Thirsty: Global Geopolitics & Political Economy/IPS. 22 March 2011.
23. Forero J. (2005). Bolivia regrets IMF experiment. *The New York Times*, 14 December 2005.
24. Segerfeldt F. (2005). Water For Sale: How Business and the Market Can Resolve the World's Water Crisis: Cato Institute 2005. ISBN: 1-930865-76-7.
25. Celebrating community management of water in Cochabamba: E-Source News, 11 May 2011.
26. 'Social Control' and Public-Collective Partnerships with Community-Run Systems in Cochabamba, Bolivia: Case Study 5, Our Water Commons. (accessed 2013)
27. PSP in Bolivia and Argentina – What Actually Happened? The Sound of Thirst: Why urban water for all is essential, achievable and affordable. Parthian Books 2012.
28. Towards sustainable water services in La Paz, Bolivia – IRC: www.irc.nl/page/76730 (accessed July 2013)

29. Feil J. (2003). Observations on the National Competition Policy Water Reform Program. Urban Water 2003. Advancing Sustainable Development. Quorum United Conference. Melbourne, October 2003. www.ncc.gov.au (accessed July 2013)
30. Laval P. (2003). The Contribution of Outsourcing to Sustainability. Sustainable Urban Services. Seminar, Shanghai.
31. Birnbauer W. (2003). *The Big Pong Down Under*. International Consortium of Investigative Journalists, February 2003. http://www.icij.org/contributors/bill-birnbauer (accessed July 2013)
32. Mac Peter (2003). Water Privatisation the Problem, not the Solution. The Guardian. www.cpa.org.au/garchve03 (accessed 14 May 2003)
33. Grisoti M. E. and Saunders B. J. (2002). The Adelaide Outsourcing Contracts: Developing a Successful Risk Sharing Partnership with the Private Sector. Conference: Public and Private Sector Partnerships. Auckland, August 2002.
34. Spoehr J., Quiggin J., Wilson L. and Purse K. (2002). State of Secrecy – Outsourcing: Promise and Performance. Prepared for the Public Service Association of South Australia, January 2002.
35. Hamnett S. (2003). Adelaide as a Sustainable City: Water Management in the Broader Context of Sustainability. Papers to the Sustainable Cities Task Force Conference, China, April 2003.
36. Infrastructure Partnerships Australia: Case Studies. http://www.infrastructure.org.au/Content/CaseStudies.aspx (accessed July 2013)
37. Transparency Statement: Water and Wastewater Prices 2005–6. Government of South Australia.
38. WSAA Facts. www.wsaa.asn.au (accessed July 2013)
39. SA Water Annual Report 2004–5 Government of South Australia. www.sawater.com.au (accessed July 2013)
40. New Metropolitan Water and Wastewater Alliance Starts Tomorrow: Press Release Government of South Australia/SA Water. www.sawater.com.au/NR/rdonlyres/C4672F06-DDF7-43A9-97CD-C248187F2BB6/0/MedRelAllwaterAlliance300611.pdf (accessed 30 June 2011)
41. Lanti A. (2006). A regulatory approach to the Jakarta water supply concession. *International Journal of Water Resource Development*, **22**(2), 255–276.
42. Harsono A. (2005). When Water and Political Power Intersect. Nieman Reports Spring, http://www.nieman.harvard.edu/reports/article/101044/When-Water-and-Political-Power-Intersect.aspx (accessed July 2013)
43. Harsono A. (2003). Water and Politics in the Fall of Suharto. The Center for Public Integrity. www.publicintegrity.net. (accessed July 2013)
44. Jensen O. (2005). Troubled Partnerships: Problems and Coping Strategies in Jakarta's Water Concessions. 4th Conference on Infrastructure Research. Berlin, October 2005. www.wip.tu-berlin.de (accessed July 2013)
45. Walhi to Refile Petition Against Water Law (2005). National News, July 22, 2005. The Jakarta Post, Jakarta.
46. Councillors Blast Palyja for Selling Bulk of its Shares (2006). City News, July 28, 2006. The Jakarta Post, Jakarta.
47. Jokowi to cancel contracts with water firms: The Jakarta Post 28 March 2013.
48. City wants water takeover to avoid penalties: The Jakarta Post 30 March 2013.

49. Case Study Water, Jakarta Indonesia: The Special Unit for South-South Co-operation (2012). www.esc-pau.fr/ppp/documents/featured_projects/indonesia.pdf (accessed July 2013)
50. Erridge A., McIlroy J. and Fee R. (2001). Best Practice Procurement. Gower Publishing Ltd, Aldershot, UK.
51. The Orange Book (2004). Management of Risk – Principles and Concepts. HM Treasury, UK.
52. Farley M. and Trow S. (2001). Losses in Water Distribution Networks: A Practitioner's Guide to Assessment, Monitoring and Control. IWA Publishing. ISBN: 9781900222112.
53. Rising to the Challenge. Scottish Water Solutions. Annual Report and Accounts, 2004–5. www.scottishwater.co.uk (accessed July 2013)
54. Investing in Scotland's Water Industry: Improvements Delivered in 2006–10. Scottish Government. www.scotland.gov.uk/Resource/Doc/917/0112271.pdf (accessed July 2013)
55. Scottish Water Solutions 2: Building on Success (2010). Presentation at Institute of Water Autumn Seminar, 17 September 2010, Institute of Water, Tyne and Wear, UK.
56. Bitran G. A. and Valenzuela E. P. (2003). *Water Services in Chile: Comparing Private and Public Performance*. Public Policy for the Private Sector, Note no. 255. World Bank, Washington DC.
57. Carvo M. and Cariola E. (2004). Analysis of the Privatisation Process of the Water and Sanitation Sector in Chile. Draft Working Document. prepared for the UNRISD Project on Commercialisation, Privatisation and Universal Access to Water. September 2004.
58. Calvo M. A. A. and Cariola E. C. (2006). Historia del sector sanitario chileno: De la gestión estatal hasta el proceso de privatización. Programme on Social Policy and Development, Paper No. 29. United Nations Research Institute for Social Development, Geneva.
59. Shirley M. M., Xu C. L. and Zuluaga A. M. (2000). Reforming Urban Water Supply: The Case of Chile. Policy Research Working Paper, World Bank, March 2000.
60. Earthquake capex to boost Chile's water spend: Global Water Intelligence. Vol. 11, Issue 8, August 2010.
61. Informe de Gestión del Sector Sanitario (2011). Superintendencia de Servicios Sanitarios, Gobierno de Chile.
62. Fu Derin Tao, Chang Miao and Zhong Lijin (eds) (2007). Reform of China's Urban Water Sector. IWA Publishing. ISBN: 9781843391432.
63. Public-private partnerships for water supply and sanitation services in Macau (China): Good Practices Inventory, Research on Innovative and Strategic Policy Options, Asia-Pacific Environmental Innovation Strategies (APEIS). enviroscope.iges.or.jp/contents/APEIS/RISPO/p_report_1st/ap_11-20.pdf (accessed July 2013)
64. Macao Water Annual Report 2011: The Macao Water Supply Company Limited. www.macaowater.com (accessed July 2013)

Chapter 9

The story of England and Wales

A sequence of developments

The decision to privatise water was taken by the Conservative Government led by Margaret Thatcher, as part of a policy to move the provision of utility and other services from the public to the private sector. The decision was taken in the strongly held belief that private sector operations are more efficient, cut costs and that this provides the economic climate for enterprise and a strong economy. Initially consideration was given to a concession approach based on the French model, but this was discarded in favour of the equity-based model which was regarded by the Government as successful with other privatisations, such as that of British Telecom in the telecommunications sector. There have been a number of books ([1], [2], [3] and [4]) published on privatisation of water and sewerage in England and Wales which took place in 1989. Some give a favourable report and others are critical. *The Development of the Water Industry in England and Wales* [5] prepared jointly by the Department for Environment, Food, and Rural Affairs and the Office of Water Services, Ofwat, published in 2006, provides an excellent account of the history.

To understand the situation leading up to privatisation it is important to look at some earlier history. The major changes began in 1974 although important steps had taken place from 1945 onwards. The Water Act 1945 began the process of consolidation in water supply with the requirement that local authorities form joint Water Boards. The impact of this was that the water supply bodies (over 1000) in existence at that time had consolidated to 198 in 1973. The Rivers Board Act 1948 was the beginning of pollution control through river management, and the River Boards became River Authorities in the Water Resources Act 1963, with the added responsibility of conserving water resources. The 1963 Act established a national body, the Water Resources Board to advise Government on long-term water resource needs. These were important steps but to my mind the Water Act 1973 provided many of the key elements for the change to private sector ownership and management in 1989.

HISTORY OF WATER SERVICES IN ENGLAND AND WALES 1974–1989

The Development of the Water Industry in England and Wales [5] provides an authoritative account of the history, but as I was closely involved from 1974 to the present day, I give a short personal account, not only to repeat some important facts but also to try to give a flavour of both the various authorities' and the public's attitudes at each stage.

Prior to 1974 most of water supply services, and all of sewerage services, were managed by local authorities. There were 1620 public service water bodies in existence at that time; 198 local government organisations and Joint Water Boards, 1393 sewage and sewerage authorities, and 29 river authorities. Some of the "Water Boards" were large like the Metropolitan Water Board serving London, which itself had been formed from 9 separate units in 1903, and there were some small local authority operations. There were also 33 statutory private water companies, some of which, as mentioned in Chapter 8, had been formed in the middle of the 19th Century. The Water Act of 1973 made provision for the formation of 10 Regional Water Authorities (RWAs), structured on river basins with their boundaries being the watershed lines separating river systems. All of the Joint Water Boards and the local authority units, together with the river authorities, were incorporated into RWAs based on their locations within the river basins, although there were some anomalies between water and sewerage due to historic service provisions. At that time the Water Resources Board was abolished but another national body, the National Water Council was formed. The formation of the RWAs has been described as a process of nationalisation, although it was different from the immediate post World War II changes in the UK, when the Labour Government of the time took utility and other industries, such as gas, electricity, mines, steel and transport, into public ownership. It was nationalisation in the sense that locally-controlled services became national entities directly responsible to national Government, although for a time local authority representation controlled the boards of the RWAs. There was great resistance from some of the local authorities to losing their utilities and legal action was taken against central Government to try to prevent the transfer of assets. These actions failed. In some cases the resistance was evident in an initial period of non co-operation with the new RWAs, who were denied access to records. I remember at the time a report of one RWA, on the weekend before it took over responsibility for the services, having to advertise in local newspapers to invite staff, who had been working for the local authority-run water utility, to turn up on the Monday morning to give their names and take up the continuation of their jobs. Unfortunately I didn't obtain a copy of the newspaper advertisement at the time and am now unable to trace it, but this story illustrates the level of resistance to the changes. Some of the inherited systems were in good condition, and some were in a neglected state. In the North West Water area, RWA staff had been unable to prepare a complete inventory of all the sewage treatment plants, as some small percolating filters had been forgotten and were not found until later, hidden behind overgrown vegetation.

The need for action on sewage treatment had been highlighted earlier in *Taken for Granted*, the 1970 report of the Working Party on Sewage Disposal [6], chaired by Lena Jeger, MP, set up by the ministry then responsible for water, the Ministry of Housing and Local Government. The "Jeger" report, and the poor quality inherited infrastructure discovered by the RWAs vindicated the decision to integrate many small utilities into viable sized units. However, these changes resulted in the loss of local identity and civic pride which had produced some world leading operations at the time. I return to this later.

The argument for river basin authorities was not only viable utility size but also the need to manage river systems in an integrated way, both to optimise investment in water and sewage treatment and to achieve the desired environmental improvements in river water quality. The formation of RWAs in 1974 was heralded [7] as the way forward in water management. River basin management systems had been in place earlier elsewhere in Europe, one of the best known examples being the Ruhrverband (Ruhr River Association) which was created at the beginning of the 20th Century, but the formation of the RWAs was the first time that river management, water supply and wastewater services had been integrated on a large national scale.

One other major change in 1974 was the cessation of government grants for infrastructure development, and from that time onwards water services had to be self-financing. In some areas, water service charges had been included in local taxation, so it was a new experience for many people to receive water bills. Until that time, consumers had not experienced the cost of their water services. It is not clear that all local authorities reduced the level of local taxation by the appropriate amount and many people saw a significant increase in their total service payments. So the introduction of the RWAs became controversial, with media attacks on such aspects as senior staff salaries and the cost of RWA head offices. One particular but ill-advised action that achieved headlines was the purchase of a car number plate, reading RWA 1, by the Chairman of North West Water Authority. The publicity forced the removal of the number plate which I was told was sold at a profit several years later. This is an example of the unpopularity of the RWAs at that time, and in my recollection, the RWAs received more adverse media coverage than privatisation 14 years later.

Although the formation of the RWAs was regarded as a major step forward in the management of river systems, the RWA regulatory role came under criticism. Manufacturing industry was concerned that its water abstractions and discharges were being regulated by RWAs who themselves had a major impact on river quality due to sewage discharges. The term used to describe this at the time was that the RWAs were both "poachers and gamekeepers". Clearly there could be conflicts of interest and it is not desirable for an operator also to be a regulator. This anomaly was not addressed until the time of privatisation.

There was Government concern over water charges at that time, especially in some areas of the country, and the Labour Government in 1977 introduced the Water Charges Equalisation Act to provide for more equal charging across the country. This involved RWAs with lower charges having to increase charges to subsidise services in higher cost areas. This was criticised by many at the time, on the basis that it gave no incentive for the RWAs to become more efficient, in fact quite the reverse. The cross subsidies were suspended by the Conservative Government in 1979.

The formation of the RWAs should have provided for a period of advancement with improved water services, investment for improved quality of drinking water, for infrastructure refurbishment, and improvements to the environment. However, investment levels actually fell after 1974 due to a combination of debt inherited from the local authorities and a period of high inflation. The inherited debt in 2003–4 prices, according to Ofwat, was £13.6 billion. The investment figures for the period 1974–1989 are given in Figure 9.1.

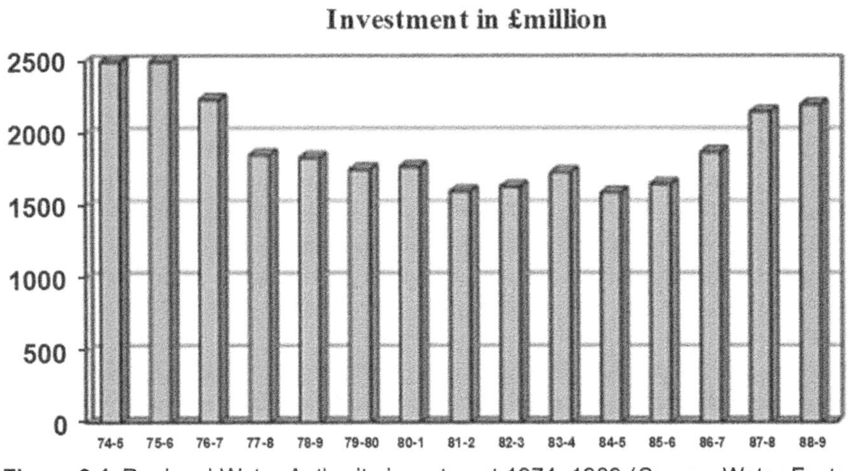

Figure 9.1 Regional Water Authority investment 1974–1989 (*Source*: Water Facts, Water UK).

It is likely that in the period leading up to the formation of the RWAs the local authorities were keen to invest in improvements in water services, particularly as they knew that the debt would not fall on them but on the RWAs.

The RWAs inherited a substantial investment programme, as can be seen by the numbers for 1974–5 and 1975–6 in Figure 9.1, but the reduced levels of investment after that coincided with the increasing problems of river pollution, dirty beaches from inadequate sewage sea outfalls, and the burden of deteriorating sewers and water mains. Sewage on beaches became a major issue in the South West of England. I can remember reading newspaper articles in which local authorities, who a few years previously had been responsible for sewage handling, complained about the effect on the tourist trade. The deteriorating water mains resulted in rusty water, a problem which was to become worse and require a high level of investment by the private companies during the period 1990–2005. Collapsing deep sewers hit the newspaper headlines with the measurement of the size of the resulting holes in the road being given a new and unique unit, the ddb. For ddb read "double-decker bus". Newspapers referred to a 1 ddb or a 3 ddb hole in the road.

One result of concern about the underground infrastructure by both Government and the RWAs was the establishment of a national committee to study sewers and water mains. The Committee's report, *A National Assessment* [8] published in June 1977, concluded that the situation had changed from expansion of systems to renewal, that there was a lack of information on asset condition, and that research was required on asset management and renewal. This report stimulated increased interest, and it was followed by a House of Lords Select Committee report [9] in 1982, which stressed the need for greater investment in renewal of the underground infrastructure. *The National Water Council Review* [10] published earlier that year, stated that annual investment in sewers and water mains needed to increase by 40% to £355 million (£970 million in 2007 prices).

At the end of the 1970s, a national technical working group was trying to obtain reliable estimates of the levels of leakage from water distribution systems in England and Wales. Although the estimates at that time were based on poor methodology, the Working Group's reports of 1976 and 1980 [11] provided a sound basis for the development of leakage control policies. The recommended practical approaches to leakage measurement and detection provided the foundation for the systems used today. Leakage became an important subject again in 2006 due to the drought and public reaction to leaks.

As a result of the concerns expressed in *A National Assessment*, the RWAs, the Statutory Water Companies, and the Department of the Environment, together with the Regional Water Authorities in Scotland and the water service operation in Northern Ireland, supported the Water Research Centre in setting up a new laboratory in Swindon. A team of researchers who had carried out some initial work on asset management in the late 1970s, moved into temporary premises in Swindon in 1980, and into a new laboratory in 1983. The research work was of a practical nature providing support to the water industry on both policy and engineering. The early results came together in two manuals. The WRC Sewerage Rehabilitation Manual was first published in 1984. It became the standard document worldwide for those concerned with the maintenance of sewerage infrastructure with three updates, the 4th Edition being published in

2001. The equivalent manual for water distribution systems, the Water Mains Rehabilitation Manual was first published in 1989. These were major achievements, not only for the research organisation, but for the Water Industry as a whole, as all the water service operators pooled their resources to tackle these major problems. This work was also important because it provided the framework for the development of the approach to asset management in the late 1980s, which was vitally important in giving a secure base for preparing the first privatisation Asset Management Plans in 1989, the so-called AMP 1. It is ironic that, although privatisation benefited from the collectively-funded research effort, it also greatly diminished the willingness of the industry to do collaborative research. There has been ongoing research through UKWIR, the UK utilities joint research organisation, which is extremely valuable, but I doubt whether the industry is in a position to consider major strategic research work on the scale that occurred on asset maintenance. The House of Lords Science and Technology Committee, in its 2006 Water Management Report [26], referred to the low level of spending on research, suggesting that a certain proportion of water companies' turnover, to be used for research, should be free from efficiency targets.

The United Kingdom had become a member of the European Union, the EU, (but at that time referred to as the European Community) in 1973 just before the formation of the RWAs, and as such became committed to complying with European Directives. Complying with these Directives added considerably to the required level of investment in water services. One important Directive was the EC Drinking Water Directive (EC80/778/EEC) which specified the minimum standards for drinking water across all Member States. The UK, being relatively new to the operation of the EU at that time, had not been able to have much influence on the details in the Directive. Andrew Semple [12], Under Secretary in the Department of the Environment during the period 1976–1983, described the UK Government's attitude to Europe and environmental legislation at that time as "Euro-agnostic, combined with a failure to comprehend what Europe was all about".

One controversial aspect in the Drinking Water Directive was the standard set for all pesticides, irrespective of the risk to health of any particular substance. The standard had not been determined on a scientific basis but on a political view that there should not be any pesticides in drinking water. The standard of 0.1 μg/l (1 part in 10 billion) was at the detection limit for organic substances at that time, and as such was a surrogate zero. The UK argument within the EU at that time was that this would result in unnecessary costs, and the money required could be spent more usefully on other things. My view was that whether or not the standard had any scientific merit didn't really matter, because the treatment necessary would provide benefits in improving the aesthetic quality of tap water through the removal of other organics affecting the taste of water. It would also add another barrier in the protection of water quality during emergencies. In any case, the cost of additional treatment would have a relatively small impact on the cost of water supply, compared with the required investment in the distribution systems.

The concern over costs delayed the implementation of the Directive. The Directive was not transposed into law in England and Wales until privatisation in 1989. It is worth noting that until the drinking water quality regulations of 1989 [13], there were no statutory numerical standards for drinking water. Thus, in the early 1980s, there was a growing crisis in investment in water services. Inadequate earlier investment in the refurbishment of the infrastructure had created a backlog of refurbishment work. This, coupled with insufficient attention having been given to the environment, and the need to comply with EU Directives resulted in a growing need for increased investment in water services, but this was not to happen until privatization. However, in 1982 there was still one more phase in the changes to the public sector structure, which was important in creating water service organisations able to transfer to the private sector.

The majority of members of the Boards of the RWAs over the period 1974–1983 were elected local authority councillors. This had the advantage of maintaining some city involvement from peoples' elected representatives but the people involved, however dedicated they may have been, could not be expected to give priority to their RWA roles. Also it might have been expected that one of their main concerns would be for the services in their own areas, rather than for the overall function of their RWA. Some of the Boards were very big, for example the Thames RWA Board had 60 members, and this did not provide for dynamic leadership. The Water Act 1983 made provision for smaller RWA Boards through a reduction in local authority councillor membership. The aim was for a more commercial approach with the appointed Chairs of the RWAs having a commercial rather than a public service background. The old style Board meetings had been held in public, but the 1983 Act removed the requirement for public attendance. To counter the unpopularity of this change, Consumer Consultative Committees were formed, one for each RWA area. Also, the last national body with any policy role, the National Water Council, was abolished. For coordination, the RWAs formed an association, the Water Authorities Association, the WAA.

The changes in the 1983 Act did create a major shift in approach with the more commercial boards introducing a drive for efficiency. Over the period 1983–1989 there were some reductions in operating costs with substantial cuts in staff numbers in some authorities. During the period 1974–1982 the operating costs of all the RWAs combined (from figures taken from Water Facts, Water UK Publications) had increased by around 27%, but they reduced by around 6% over the period 1983–1989. Following the 1983 changes, as can be seen in Figure 9.1, there was an increase in total capital expenditure across England and Wales. One intention of the 1983 Act was to increase the level of investment by allowing the RWAs to borrow money in the private money markets. However, in practice, the amount borrowed was not at the level required to meet investment needs. This was due to Government Treasury rules, which limited the total amount of borrowing by the public sector, whatever the source of funds.

The drive by some of the RWAs to cut costs resulted in substantial changes in staff numbers. A paper by Green [14] suggests that total staff numbers for the 10 RWAs reduced from 80,000 in 1974 to 50,000 in 1989. As a result of these and other cuts in operating costs, some RWAs, such as Thames, were outperforming on annual profits. The Thames RWA Board wished to pass on the benefits to its customers in reduced water charges. However, the Authority had certain debts on Government loans, which it was repaying under agreed terms. As I recall from discussions with RWA executives at the time, the Government (The Treasury) decided that rather than reduce water charges, it would require Thames Water Authority to use the operating surpluses to pay back the loans at a faster rate. The Authority Board saw this as undue Government interference in its affairs, and lobbied Members of Parliament from constituencies in the Authority's area, together with relevant Members of the House of Lords, to obtain support against the Treasury's decision. It was my impression that the Government, under pressure from this influential group, responded that it intended to privatise the RWAs. What is not known is whether Government, without this pressure, would have privatised water at any time in the future, or whether the pressure merely brought forward the timing.

Some good progress was made during the life of the RWAs in bringing the condition of those areas of the RWAs with poor quality inherited treatment facilities to a better standard. However, due to the initial cumbersome Board structures, and various restrictions on water charging levels and on the level of investment throughout the 14-year period, I think it is doubtful that the results met the expectations of the architects of the 1974 restructuring, or of those who enthused about the changes at that time. It became clear in the early 1990s that the underground structures were still in a neglected state and needed a high level of investment for refurbishment.

Whatever the true situation behind the decision, privatisation was brought about by the Water Act 1989 [15].

THE WATER ACT 1989

The Water Act 1989 made provision for the following:

- The 10 RWAs, nine in England, and one in Wales, would be privatised through flotation on the stockmarket with the aim of shares being widely owned. Special provision was made for staff and customers of RWAs to obtain shares;
- The existing 26 Statutory Water Companies would change status to plcs (public limited companies) and would be regulated in the same way as the 10 ex RWAs;

- The regulatory responsibility of the RWAs, together with the responsibility for flood control, would be transferred to a newly formed body, the National Rivers Authority. Staff would be transferred appropriately;
- Economic regulation would be independent from Government and handled by a new body, the Office for Water Services, Ofwat;
- Powers for the Secretary of State to make regulations to define the "wholesomeness" of drinking water. This resulted in the Water Supply (Water Quality) Regulations 1989;
- A Drinking Water Inspectorate, based within the Department of the Environment, would be formed to enforce drinking water quality regulations;
- The establishment of Customer Service Committees.

The full provisions can be seen in the 1989 Act and were later part of the consolidation of water legislation in the Water Industry Act 1991 [16] which was "An act to consolidate enactments to the supply of water and the provision of sewerage services, with amendments to give effect to the recommendations of the Law Commission". On drinking water, there were some very important powers given in these Acts for both enforcement of regulations and prosecution for certain offences. It was and is a tough regulatory regime. According to the allocation of powers in the Act, both the Secretary of State (i.e. the Government Minister responsible for Water) and the Director General, Ofwat, could issue an enforcement order if a water company contravened any condition of its appointment licence or any statutory requirement. The powers for action in Wales were transferred to the Welsh Assembly on its establishment in 1999. Although the powers initially rested with the Secretary of State and the Welsh Assembly, in practice enforcement of drinking water quality regulations was handled by the Drinking Water Inspectorate. The Water Act 2003 [17] recognised this delegation of powers with certain powers now being vested directly in the Chief Inspector of the Drinking Water Inspectorate. The enforcement powers were to prove highly effective in securing compliance with drinking water quality standards. Unfortunately, at that time, the National Rivers Authority did not have equivalent enforcement powers for compliance with conditions for discharges to rivers, having instead to prosecute water companies for pollution of receiving waters. The high number of prosecutions during the early years generated a lot of adverse publicity for the water companies. In many cases, the companies were dealing with problems they had inherited from the RWAs. The need for the Environment Agency (the National Rivers Authority became part of the Environment Agency in 1995) to have enforcement powers on discharges was eventually remedied in the Water Act 2003. Had those enforcement provisions been included in the 1989 Act, there would have been a much more effective mechanism for including the environmental investment needs in the 5 year periodic reviews.

Regulation of drinking water quality

On drinking water quality, a breach of the drinking water regulations results in enforcement action being taken by the Drinking Water Inspectorate, but instead of issuing an enforcement order at the initial stage, the Inspectorate invited a water company to offer an undertaking to take the necessary remedial action by a given date. Such an undertaking was legally binding and failure to meet the agreed action or timetable resulted in an enforcement order (see Sections 18–22 of the Water Industry Act 1991 for the undertaking/enforcement provisions).

This proved to be a very efficient and effective way of taking enforcement action as the necessary commitments to make the required improvements could be obtained from water companies within a few weeks, whereas any process involving Courts of Law is slow and cumbersome. Strangely, the European Commission, acting presumably on a complaint from some source, took the United Kingdom to the European Court alleging that the UK had failed to implement the Drinking Water Directive correctly in relation to arrangements for enforcement. In April 1999, after a number of years of legal wrangling, the European Court ruled that "accepting undertakings in the absence of a specific legal framework governing the conditions under which undertakings may be accepted did not constitute adequate compliance with the law with the current EC Drinking Water Directive (80/778/EEC)". Responding to this ruling required some changes to the approach [18], which although complying with the "letter" of the Directive, added nothing to the effectiveness of the enforcement process, which had achieved virtually 100% compliance with the UK regulations. The compliance results are shown in Figure 9.2, which shows the compliance percentage of around 3 million tests each year, and in Figure 9.3, which shows the number of tests not achieving compliance.

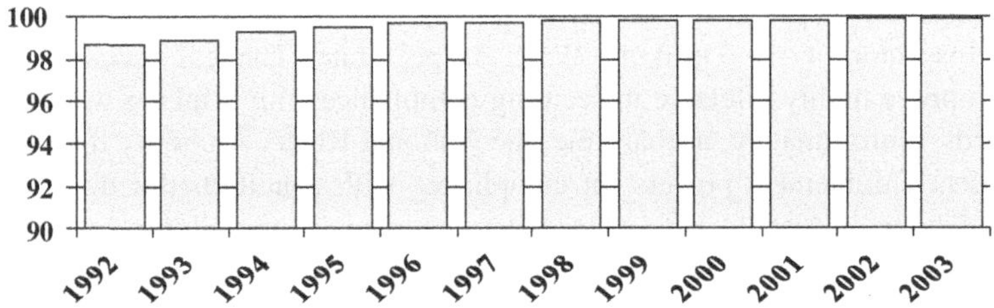

Figure 9.2 Percentage of tests complying with Drinking Water Quality Standards in England and Wales over the period 1992–2003 (*Source*: Drinking Water Inspectorate Annual Reports).

I have no doubt that the European Court interpreted the legal requirements correctly but it was a great pity that the legal proceedings were thought necessary. In my view, the proceedings were not a good reflection on a European system, which had delivered significant benefits through the introduction of geneRally sound drinking water quality requirements. Fortunately, that is now history and there is now general recognition that the UK does take drinking water quality very seriously. There was good co-operation between Member States and the European Commission in the development of the Drinking Water Directive (98/83/EC), which was transposed through the Water Supply (Water Quality) Regulations 2000 and 2001 [19]. There is ongoing co-operation through a network of European drinking water regulators (ENDWARE).

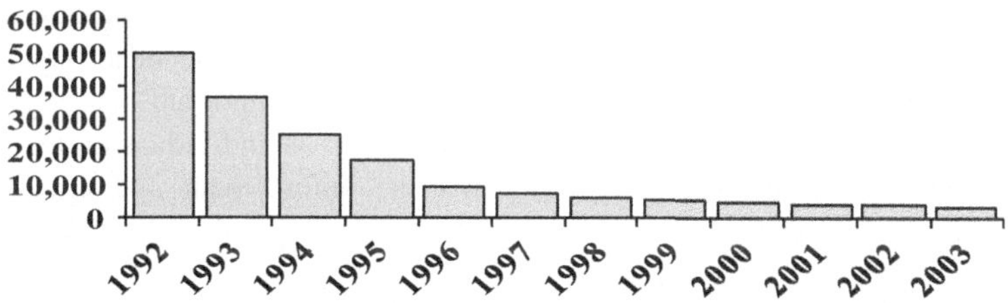

Figure 9.3 Number of tests breaching the Drinking Water Quality Standards over the Period 1992–2003 (*Source*: Drinking Water Inspectorate Annual Reports).

The 2000 Regulations changed the "baseline" for compliance. The overall compliance as a percentage of tests for the period 2004–2011 is given in Table 9.1, showing ongoing almost complete compliance. Testing remains as a regulatory requirement for compliance, with an ongoing requirement for reporting within the European Union. However, in line with WHO guidance, the emphasis has changed to risk management with water companies being required to satisfy that effective risk assessment has been carried out. Risk assessment and management are discussed in Chapter 4.

Table 9.1 Drinking water quality compliance with regulations.

Year	2004	2005	2006	2007	2008	2009	2010	2010
Compliance %	99.4	99.6	99.6	99.6	99.6	99.5	99.6	99.6

Source: Drinking Water Inspectorate Annual Reports.

If a water company supplies water that is "unfit for human consumption", it may be prosecuted under Section 70 of the 1991 Act. Water unfit for human consumption is not defined in the Act, and it is for a Court to decide on a case-by-case basis. In practice, it can be either a water that results in illness, or water that is unpalatable for aesthetic reasons of taste, smell or appearance. For example, very rusty water has been regarded in Court as unfit for human consumption. In their defence water companies have to show that they "had no reasonable grounds for suspecting that the water would be used for human consumption", or that they "took all reasonable steps and exercised all due diligence" for securing that the water was fit for human consumption on leaving its pipes, or that it was not used for human consumption'. Such prosecutions by the Drinking Water Inspectorate are infrequent. The total number of prosecution cases can be numbered in tens, whereas enforcement actions can be measured in thousands. In the early period of privatisation, the media, not understanding the enforcement process, criticised the Inspectorate for not prosecuting water companies, making comparisons with the National Rivers Authority. The enforcement process enables more rapid improvements, to the benefit of water consumers. The prosecution powers are available as an important deterrent, with Courts able to impose a fine against a company or an individual or to give a prison sentence of up to 2 years. Together the enforcement and prosecution powers provide for a very effective regulatory regime to protect consumers.

Environmental regulation

The Water Act 1989 transferred responsibility for environmental quality regulation from the RWAs to the National Rivers Authority, and later to the Environment Agency (EA) on its formation in 1995 [20]. The powers in the 1995 Act relate in general terms to protecting or enhancing the environment and to attaining the objective of sustainable development. However, there is reference to taking into account any likely costs, with the implication that environmental improvements should not be pursued at any cost. The powers related to pollution control and water resources given to the EA were equivalent to those of the National Rivers Authority. However, there is a particular note in relation to water resources which reads "nothing in this subsection shall be construed as relieving any water undertaker of the obligation to develop water resources for the purpose of performing any duty imposed on it by virtue of section 37 of the Water Industry Act 1991 (general duty to maintain water supply system)". I return to this later in connection with the 2006 drought in the southeast of England.

It is possible, based on the General Quality Assessment Scheme (GQA) [21], to look at progress that was made in improving the quality of rivers over the period 1990–2004. It is important to include this scheme even though it was superseded by the requirements of the EU Water Framework Directive discussed later in this section. There were four bases of assessment, namely biology, chemistry, nutrients and aesthetics:

- Biology – 6,500 sites are sampled twice during survey years;
- Chemistry and nutrients – around 7,000 river and canal sites are sampled 12 times each year;
- Aesthetics – 452 sites were sampled in the year 2000.

On chemistry and biology there were six classifications (A–F) for rivers, A being very good, B good, C fairly good, D fair, E poor and F bad. For general reporting purposes there are four groups: good (A and B), fair (C and D), poor, and bad. A target was set for England of 91% of total length of rivers and canals to achieve good or fair level by 2006. As can be seen in Table 9.2, this was accomplished by 2004. The quality of rivers in Wales was higher than in England in 1990 and there have been further improvements. Improvements in river quality are not dependent on the quality of sewage discharges alone, but the water companies spent around £6 billion on improvements to discharges to rivers over the period 1990–2005.

Table 9.2 Comparison of chemical condition of rivers and canals in England and Wales in 1990 and 2004.

Chemical Condition	England		Wales	
	1990 (%)	2004 (%)	1990 (%)	2004 (%)
Good	43	62	86	94
Fair	40	31	11	4
Poor	14	7	2	2
Bad	3	1	1	0

Source: Environment Facts and Figures. Environment Agency.

Also there was a significant reduction in the number of major or significant water pollution incidents since 1995, although the graph (Figure 9.4) shows a slight upward trend over the period 2000–2003. Again, this is pollution from all sources and not simply related to pollution incidents resulting from failures by the water companies.

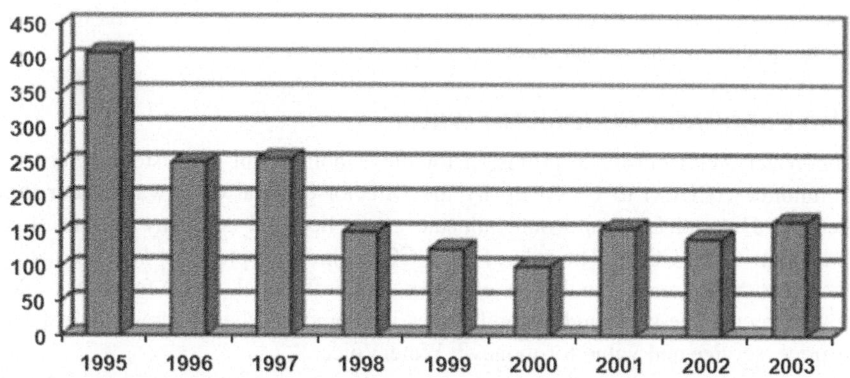

Figure 9.4 Number of major or significant water pollution incidents in England and Wales 1995–2003 (*Source*: Water UK).

In 2003 the European Water Framework Directive (WFD) [22] was transposed into UK law [23]. An introduction to the WFD [24] is given on the European Commission website. Essentially the Directive requires EU Member States to take action to achieve "good" status of all surface waters (rivers and lakes) and groundwaters. The WFD also covers estuaries and coastal waters. The timescale for implementation of the WFD is through three planning cycles, those of 2015, 2021 and 2027. The European Commission monitors progress and issues reports every two to three years. The third report [25] was adopted on 24 November 2012.

One key element of the Directive is coordination of administrative arrangements within river basin districts (Article 3). Although generally the UK, having managed the water environment through river basins, was well-placed to implement this article, it was a devolved government matter requiring arrangements to handle those rivers crossing boundaries between England and Wales, and England and Scotland. An outline of the transposition elements is given in a Defra paper [26]. In England and Wales, the EA has responsibility for drawing up river basin management plans and for monitoring environmental water quality. The quality measures are different and more demanding than the GQA Scheme. It is based on a far wider range of assessments than the GQA classification. There are over 30 measures, grouped into ecological status (including biology and "elements" such as phosphorus and pH) and chemical status ("priority substances"). It is difficult to compare current data with that published earlier. The EA quality report [27] appears to show either no progress or some deterioration in overall surface waters quality over the period 2009–2012. However, the EA report has a "data health" warning which reads "Please note that simple comparison with the 2009 baseline results is not possible as differences could be due to methodological changes, for instance using data from a greater number of quality indicators, rather than genuine environmental change. A true assessment of progress will be made in 2015/6 when the current river basin cycle is complete." The change in monitoring from the GQA scheme does not allow meaningful trends to be made at the time of writing. For this reason I am not including any of that data.

The consumer council for water

The Water Act 1989 made provision for the establishment of Customer Service Committees (CCCs) to be set up by the Director-General of Ofwat. Under the Act, the Director-General would allocate every licenced company to one of a maximum of 10 CCCs. The role of the CCCs, which adopted the working name of Water Voice, was "to be an effective and influential voice for water and sewerage customers in England and Wales in promoting their interests in respect of price, service and value for money". WaterVoice:

- Dealt with customer complaints which had not been dealt with successfully by companies;

- Sought compensation from companies who had failed to provide the required service;
- Was the main voice of customers during Periodic Reviews which are discussed in a following section.

WaterVoice was very effective in handling complaints and seeking compensation for consumers, and worked very closely with the Drinking Water Inspectorate on drinking water quality complaints. However, it was part of Ofwat and, although it became more independent over time, the Water Act 2003 made provision for the replacement of the Customer Service Committees by a wholly independent body, the Consumer Council for Water (CCWater). The Council came into existence in October 2005 and took over the duties and staff of the previous Committees. It is a national body with regional offices. It aims to be a strong national voice for water consumers and on its formation set five priorities:

- Value for money: a fair affordable price and charging system;
- Right first time: problems sorted out quickly without hassle;
- Water on tap: a safe, secure and reliable supply of water, used wisely;
- Flush and go: responsible removal of sewage, prevention of sewer flooding, and reduction in persistent smells from sewage treatment works;
- Speaking up for water consumers: achieving improvements.

CCWater produces annual reports [28] giving its performance on, for example, handling complaints. Table 9.3 gives the performance results on two aspects from the 2011–12 report. CCW sets targets for improvement and generally they have been exceeded. Progress requires water companies to respond to CCW pressure.

Table 9.3 Examples of CCWater performance over the period 2006–7 to 2011–12.

Year	2006–7	2007–8	2008–9	2009–10	2010–11	2011–12
Customers satisfied with the Quality of CCWater service (%)	61	63	67	74	75	75
Complaints resolved within 20 days (%)	72	76	76	82	82.1	83.9

Source: CCWater Annual Report 2011–12 [28].

CCW also produces forward plans [29]. In the plan for 2013–16 it lists its achievements since it was established in 2005. For a number of the achievements CCWater uses the words "contributed to" as recognition that it is part of a governance structure. The list includes the following:

- contributed to the 2009 price review being better for customers by the equivalent of £50 per customer compared with the 2004 review;

- £15 million pounds returned to domestic and business customers in compensation and rebates since 2005;
- Handled 290,000 complaints and enquiries about water companies since 2005;
- 40% reduction on the number of complaints about water companies;
- 73% increase since 2008 in the number of customers registered on water companies "special assistance registers" through greater awareness promoted by CCWater;
- A more cost-efficient CCWater with cost per customer in 2013–14 to be 21p compared with 24p in 2008–9.

An important activity has been promoting "Water Sure" which is a scheme to provide assistance to the vulnerable in the payment of bills. CCWater rebranded this scheme and introduced a single simple form for claimants. It provides for lower payments for those who have specific medical conditions or/and are receiving benefits. A simple flow diagram advises consumers on their eligibility for Water Sure. Almost 4 times as many customers have signed up for assistance since rebranding.

The forward programme has been based on what water consumers told them they wanted. In listing key themes in the framework of the work, CCWater stresses that the following can be achieved only by working closely with governments, water companies and others.

- Influencing the 2014 period price review in consumers' interests.
- Helping customers who are struggling to pay. Interestingly CCWater recognises that the help to those who say that their bills are unaffordable has to be acceptable to those customers who fund that assistance through the bills they pay.
- Helping business customers, specifically in helping small to medium businesses in decisions on possible switching to other suppliers. The increase in competition is discussed in a section (Ongoing Developments) later in this chapter.
- Providing information and advice to customers.
- Influencing water resource planning. 82% of customers give safe reliable clean drinking water as the top priority.
- Pressing water companies to get their customer service right first time and reduce complaints.

In a review [30] of Ofwat and consumer representation published in 2011, it was recommended and I quote "The review team is strongly of the view that effective consumer representation will be required in the sector for the foreseeable future and that the functions currently undertaken by CCWater should be maintained and protected in any new approach. Indeed, we can see areas in which the consumer role could be extended." CCWater was seen to have an important role in the introduction of social tariffs. At the time of writing the Water Companies Social Tariffs Bill is before Parliament. I return to the review reports' conclusions on Ofwat in a later section in this chapter on "Ongoing Developments".

Standards of service performance

Customer service performance is regulated by Ofwat through a Guaranteed Standards Scheme (GSS) [31]. The guaranteed standards of service are laid down by the Government. The scheme sets out the standards and the conditions under which water and sewerage customers are entitled to compensation. The scheme includes the following:

- Making and keeping appointments;
- Responding to account queries;
- Responding to complaints;
- Interruptions to supply, both planned and unplanned;
- Flooding from sewers;
- Low pressure of water supply.

In April 2008 the Government introduced revised GSS Regulations. The 2008 GSS Regulations extended provisions to include a minimum payment for internal sewer flooding, and to include external sewer flooding. The revised Regulations also consolidated all previous amendments.

An example of a compensation payment is £25 minimum for low pressure. Although not a part of the guaranteed standards scheme, water companies have agreed to a change in their licences to include loss of supply in the event of a drought. Compensation would be paid for loss of water supplies for purposes such as cooking, washing, drinking and flushing the toilet. It would not include uses such as garden watering, washing a car or filling a pool.

Overall, based on Ofwat's annual levels of service reports [31, 32] there have been significant improvements in performance. The detailed results for each water company can be seen in these reports. Two examples taken from the reports are illustrated here. On pressure of supply the levels of improvement for England and Wales in total can be seen in Figures 9.5 and Table 9.4. Incidents of low pressure in 2009–10 were virtually zero. On flooding of properties by sewage (Figure 9.6) the situation is less clear. The number of properties flooded in 2008–9 and 2009–10 are given as 5700 and 6200 which are increases over the years 2001–2 to 2003–4.The main cause given for the increase is sewer overloading. Flooding is partly dependent on the number of severe storms in each year and this is variable, and there does appear to be an increase in extreme rainfall events. There have been significant improvements in drinking water quality, in the environment and in customer service. How much investment was required, how was it financed and what was the planning process?

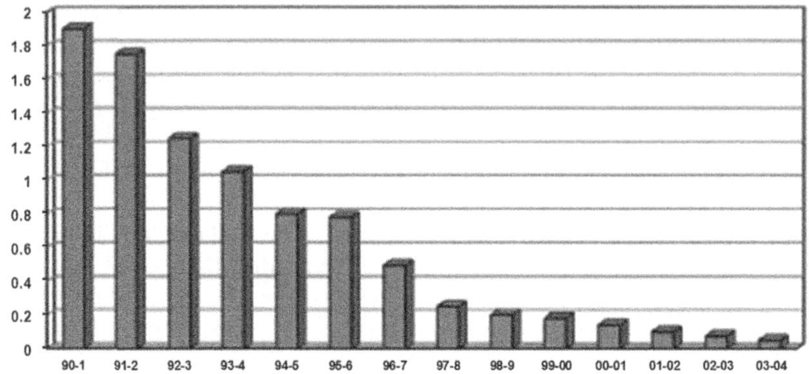

Figure 9.5 Percentage of water supplies at low pressure in England and Wales for the period 1990–1 to 2003–4 (*Source*: Ofwat Levels of Service Report 2003–4).

Table 9.4 Percentage of water supplies at low pressure in England and Wales for the period 2006–7 to 2009–10.

Year	2005–6	2006–7	2007–8	2008–9	2009–10
Percentage of water supplies at low pressure in England and Wales	0.03	0.02	0.02	0.03	0.01

Source: Ofwat Levels of Service Report 2009–10.

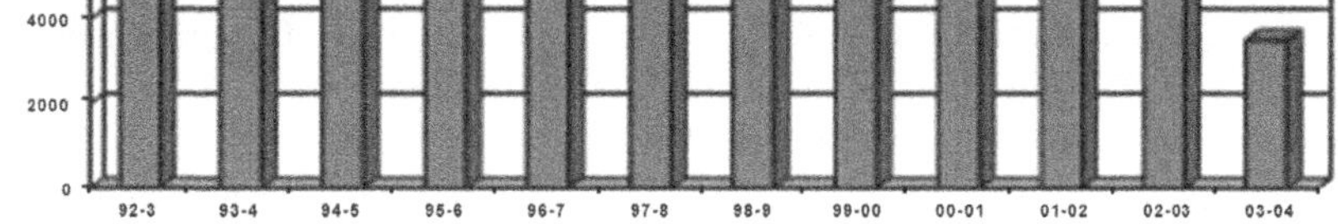

Figure 9.6 Number of properties flooded with sewage over the period 1992–3 to 2003–4 (*Source*: Ofwat Levels of Service Report 2003–4).

The periodic review process

This is the process for setting allowable water service charges in a five-year cycle, but the process also plays a pivotal role in providing an integrated approach to future planning of water services, and in the maintenance of the water and sewerage infrastructure. The process is described in full in the Ofwat publication, *Setting water and sewerage price limits for 2005–10: Framework and Approach*, where a detailed account of Ofwat's role can be found [33]. Over time, details of regulatory processes evolve, but the simplified version of the periodic review process described here gives an outline of the sequence of steps which lead to the determination of the price limits for each five-year period.

(1) Ofwat, the economic regulator, publishes the timetable of events.
(2) Customers are consulted through various market research exercises, some at the beginning and some later when Ofwat has produced its interim determinations. The surveys are organised jointly by Government, the regulators, the Consumer Council for Water, and the water companies and by the individual stakeholders, depending on the scope of the surveys. The aim is to gain a better understanding on which improvements customers would like to see and how much extra, if any, they are prepared to pay for them.
(3) Government publishes its objectives including, in outline, the service improvements it would like to see achieved during that period.
(4) DWI (the Drinking Water Inspectorate) and the EA (Environment Agency) then prepare guidance for the water companies on what is required to achieve any desired quality improvements.
(5) The water companies then liaise with DWI and the EA on how they will meet the requirements, and the two quality regulators say whether the proposals are appropriate.
(6) The water companies then submit business plans giving the improvement schemes and their costs.
(7) Ofwat considers these plans and whether the costs are realistic and, taking into account both the costs and the required efficiency improvements, calculates the required water prices to pay for the work whilst giving the companies a reasonable profit. These are given as price limits, which are the maximum average water charges allowed to be applied by water companies.
(8) Reporters assist Ofwat in the submission evaluation process. Reporters are independent professionals who, although appointed by the water companies, have a primary duty of care to Ofwat. [34]
(9) These price limits are then reviewed, including by Government, who consider whether the quality improvements are affordable or whether some aspects should wait until a later period.
(10) There are then discussions between Ofwat and each company before Ofwat announces its decision on the price limits for each company.

When the price determination results in increases in water charges, criticism is directed at Ofwat for allowing increases and at the water companies for making profits. It is important to recognise that Ofwat has to achieve a balance between protecting consumers against higher charges than necessary to achieve the specified standards of delivery, and ensuring that the water companies are able to finance the operations. Ofwat makes its determination based on the investment needed by water companies to meet Government's objectives, taking into account a reasonable level of profit whilst becoming more efficient. The aim in setting the efficiency targets is to bring the poorer performers up to the level of the better ones. If a water company outperforms whilst meeting all the investment and performance requirements, then it makes additional profits on which it can decide, during the remainder of that five-year period, how to share the money between reducing customer charges and paying additional dividends to its shareholders. At the end of the five-year period all of the improved financial performance goes to the benefit of consumers in the next five-year period. If a water company has not achieved the commitments in the five-year period, Ofwat imposes financial penalties. If quality requirements have not been met, as explained earlier, DWI and the EA will have taken enforcement action.

In order to achieve drinking water quality standards and to meet environmental requirements, the investment needed resulted in an increase in average water charges over the period 1990–2007 of around 39% in real terms. Ofwat estimated [35] that the efficiency improvements made by the water companies over that period absorbed 70% of the investment cost, without which water charges would have been considerably higher. During that period, total investment was over £50 billion pounds, over twice as much in real terms as in the preceding 15 years of public sector operation. So the question should not be whether the water companies are making profits, but rather what are they delivering? Are they delivering the required standards of service, drinking water quality and required environmental improvements at a cost which is lower than that which might have been achieved by a public sector operation? This is a difficult question to answer. I have discussed good public sector performance in Chapter 7, and I discuss the relative merits of public and private in the summary, Chapter 10.

Aside from the success of the regulatory processes in delivering improvements, it is inevitable that there would be tensions between Ofwat and the water companies. The water companies had always been critical of the regulatory burden, feeling that it placed excessive demands on time and resources. In 2011, Government commissioned an independent review of Ofwat and consumer representation in the water sector. The "Gray" report [30] concluded that regulator structural changes were not necessary but made a number of recommendations. Three of the key recommendations relating to Ofwat were:

- Ofwat needs to engage more constructively and effectively with the full range of stakeholders in the sector and be more transparent in its decision making. (This was made in the context of seeing Ofwat as a rather remote body).

- Ofwat should set clear targets and timescales for a reduction in the burden of the price control and compliance processes and enter into a joint project with the industry to achieve this.
- Ofwat to seek to ensure that the future framework of incentives provides the right balance between rewards and penalties.

In chronological order, comment on the "Gray" review of Ofwat would have come later in this chapter but the purpose of the review and its recommendations relate largely to the Periodic Review processes, and it is important to note that regulation is an evolving scene. However, how these recommendations might be taken forward should be considered in the context of the major changes, which are taking place on competition within the sector. This is discussed at the end of this chapter under Ongoing Developments.

THE "DROUGHT" OF 2006 AND THE "NON-DROUGHT" OF 2012

Water had rarely hit the headlines since the drought of 1995, but water shortages and the imposition of hose-pipe bans in 2006 provided the background to media attacks on private companies with stories centred on profits and leakage. But beyond the media hype, the situation raised important questions. If the regulatory system is so good, why is there a water shortage problem in the south-east of England? Is this due to a failure of the current institutional structure?

The House of Lords Science and Technology Committee thought so. In its report [36] it refers to the challenge presented by a combination of threats to water resources and increasing demand for water. The Committee agreed with the generally accepted view that meeting this challenge requires a twin-track balanced approach of resource development and demand management. However, the conclusion is, and I quote from the report, "This balance will not be achieved until the currently fragmented institutional arrangements for water management are simplified and coordinated". The Committee made a number of recommendations of which the key ones are:

- There should be wider stakeholder engagement through the establishment of new regional boards made up of bodies covering consumer and environmental interests and Ofwat. Such boards would produce resource development and demand management plans which would guide Ofwat's funding decisions.
- Ofwat and the Environment Agency should take a realistic approach to the essential development of new resources and, to enable longer-term planning, Ofwat should agree indicative water prices with each company for up to 24 years.
- Ofwat and the Environment Agency should also work together to ensure that water companies maximise their promotion of water efficiency.

- Ofwat should sanction increased water company expenditure in reducing leakage, and the reduction targets should take account of economic, social and environmental factors.
- The Government should allow compulsory metering in the driest parts of the country.

I think that the Committee's analysis was sound, but I do wonder about the implications of some of the recommendations. During the same period of the Committee's deliberations, a student on the Oxford University MSc Course on Water Science, Policy and Management, was testing the hypothesis that the water shortage in the south-east of England was due to institutional failure. She considered the history of water resource planning from the 19th Century through to the present time, and interviewed the key stakeholders. Her conclusions [37] have some similarities to those of the House of Lords Committee. She refers to the need for a longer planning horizon in the Periodic Review, and the need for the Environment Agency to give more attention to its role in water resource development, but she places much greater emphasis on the role of Government. She says, and I quote, "Alongside the promotion of demand management and water efficiency, the need for resource development must be considered where it is necessary. Defra (Department of Environment, Food and Rural Affairs), with its responsibility for overarching policy guidance, can play a key role in this regard."

It seems to me that one important aspect, not discussed in consideration of "failures" of the regulatory system, was the impact of EU Directives on UK Government policy. Prior to 2006 the whole emphasis in EU legislation has been on quality, whether that be drinking water, or the water environment. Failure to meet the requirement of the EU Directives results in infraction proceedings, so it was understandable that much of Government's effort went into Directive handling. However, with the emphasis on quality and pollution control, water resource quantity and water security had tended to be neglected. Although there could be value in having regional boards as suggested in the House of Lords report to provide sound input to the regulatory process, I do not see the need for major changes to the institutional structure. I was concerned to see within the House of Lords Committee Report a suggestion that the proposed boards should have a direct role in the setting of price limits. This is neither necessary nor desirable as there would be the risk of price limits becoming determined on other than objective grounds. Government has the role of setting policy, which should include definition of acceptable levels of water security of supply, to reflect the public's concern over water restrictions. As part of this policy, Government should ensure that provision is made within the existing periodic review structure for appropriate planning horizons for long-term resource development, and ensure that the Environment Agency fulfils both its environment protection and water resource development roles. How water security of supply is achieved through a combination of new resources, leakage control and demand management is then a matter for water companies working with the regulatory system. I felt at the time of the House of Lords report that it was important that the deficiencies in the regulatory system highlighted by the 2006 drought were recognised, and acted on by all involved, to avoid pressures which might damage or destroy an excellent integrated planning and price setting system. In this regard it was good to note Ofwat's positive response [38] to the House of Lords Report.

Provisions for future drought management were introduced through the Drought Direction 2011 [39] and information from Defra on Drought Permits and Drought Orders 2011 [40]. A research project considered Regulatory barriers and constraints to effective interconnectivity of water supplies [41]. Extension of metering came under scrutiny in the "Walker Report" [42]. The recommendation from this report, if implemented would extend domestic metering coverage in water stressed areas to 80% by 2020.

In February 2012, following low rainfall over two winters, and with groundwater levels below that during the severe drought experienced in 1976, government announced that there was an official drought in South-East England. Water use restrictions would be introduced progressively. At the end of March 2012 the situation appeared to be serious with the prospects of severe shortages by the end of the year. In the event it started to rain in April and continued throughout the remainder of the year with 2012 declared the wettest year since recording began. Severe flooding replaced drought conditions.

ONGOING DEVELOPMENTS

Two studies commissioned by government were published in 2009, the "Walker Report" already mentioned and the "Cave Report" [43]. The Walker Report, the Independent Review of Charging for Household Water and Sewerage Services, although best known for its recommendations on domestic metering, covered a number of other important aspects. It made recommendations on water efficiency, the problem of bad debt, flooding and surface drainage, high water charges in the South West of England, and a package of help targeted on low-income customers. In the Water Industry (Finance Assistance) Act 2012 provision was made for customers in the South West of England to receive a £50 reduction in their bills commencing in April 2013. As one measure to alleviate flooding, Walker suggested a charging system which would incentivise households to minimise the surface drainage from their properties. On concerns over low income customers, there is a Bill (2013) before Parliament which would require water companies to introduce social tariffs for those customers for which the water bill would exceed 5% of income. On bad debt (ie non-payment of water bills), the Walker Report mentioned an additional £12 per year being "levied" on those customers who do pay because of the "freeloaders". I believe that freeloaders is the correct term because it is largely a matter of "won't pay" rather than can't pay. Water companies have limited redress, as cutting off supplies for non-payment is not allowed by law. I consider it wholly unacceptable that honest customers are being penalised. Government, recognising that much of the bad debt (80%) arises from non-payment in rented properties made provision in the Flood and Management Act 2010 for landlords to be jointly liable for water and sewerage bills on failure to provide details to water companies on tenants. Government issued a consultation paper [44] on bad debt in April 2012 in which it mentioned a figure of £15 impact on customers" bills, higher than in the Walker report. Although bad debt is mentioned in the Draft Water Bill 2012 [45], at the time of writing I am not aware of other measures being taken. One simple way of solving the problem would be to allow water companies to install some form of pre-payment meter or a form of flow restrictor to "encourage" payments. Provision could be made to ensure that those who genuinely couldn't pay would be protected.

The Cave Report, the Independent Review of Competition and Innovation in Water Markets, recommended a much greater market approach to water services. The report covered many aspects. On water abstraction there were four recommendations connected with problems of over-abstraction and under-utilisation of licences:

- In those areas where licence levels are unsustainable, the Environment Agency should be permitted to run reverse auctions.
- Consideration of whether simplification of Section 52 of the Water Resource Act could reduce administrative costs.
- In catchments where licence levels are sustainable, licences should be fully tradable subject only to modification for direct environmental impacts or impacts on other users from a change of use or location.
- The Environment Agency should take forward a more risk based approach to allocating abstraction licences. There is mention of setting up some form of independent contracting organisation to relieve incumbents from the supplier of last resort duty. Until such time, the Environment Agency should make use of its existing regulatory powers to reallocate abstraction licences and also issue temporary licences to facilitate upstream competition.

On discharge consents, it was recommended that "Discharge consent holders should be able to trade their discharge consents on a pollutant basis subject only to modification for direct environmental impacts from a change of location." To address the concern about the preference for capital based rather than improved operational performance solutions, the report recommended that "Companies should be given greater efficiency incentives for significant and sustained outperformance. To introduce the same symmetry and incentives to capital investment, such an approach should also be applied to operating expenditure." I do not comment on the above, but there are two aspects, one which is surprising to me, and a series of recommendations which relate to competition in the provision of water services which have very significant implications for the future of the water industry in England and Wales.

The surprising one in a report pushing market approaches relates to collaborative research and the three recommendations read as follows:

- "The UK and Welsh Assembly Governments, the industry, regulators, suppliers, the research councils and the Technology Strategy Board and other stakeholders should come together to form a national water research and development body and agree a shared research and development vision for the industry.
- To support this vision, I propose the creation of an industry research and development body. The body would be supported by funding, which could be of the order of £20 million a year, which would be allocated on a competitive basis to organisations undertaking basic research, development and trialling as appropriate. This should be in addition to current spending. The fund would be open to all organisations and water companies would be able to recover half their subscription from customers, at a cost of around 50 pence a year. Any profits from patents or licenses would be returned to members and, in the case of water companies, to customers and shareholders on an equal basis. In the first instance, the fund should be established for 10 years. Expenditure on research and development should be excluded from Ofwat's efficiency comparison tables.
- To enable Ofwat to better support innovation in the industry, for example by taking forward the research and development board and other recommendations in this report, I propose that the regulator be given a statutory duty to promote innovation. This should further the interests of customers and the environment. Ofwat should also have a statutory duty to report to the UK and Welsh Assembly Governments every five years on the measures it has taken to support innovation and how effective these measures have been."

These recommendations echo the concern of the House of Lords Science and Technology Committee in the 2006 Water Management Report [36] discussed earlier in this chapter. The Cave Report did not recommend a return to subscription funding of a single research organisation. However it showed strong recognition that much of water research is not about trying to produce saleable inventions, for which the returns are likely to be trivial compared with the operational businesses, but of water companies together addressing strategic issues which could significantly advance water and sanitation services. I had become concerned that many of the water companies were neglecting longer term strategic research and with little interest in learning from what is happening elsewhere in the world. In 2013 I see a welcome change with much stronger links developing between water companies and universities. The Cave Report recommendations, if implemented through regulatory incentives, would address the research funding deficiency. However, there was no mention of research in either the following white paper or in the draft water bill which are discussed below under the main thrust of greater competition and unbundling of licences in the Cave Report.

The key recommendations on competition related to moving from vertical integration in which water companies are responsible for water resources, water treatment, water distribution and retail services to individual licences for each component as exist in the energy industries. The report recommended legal separation of water companies' retail functions. It recommended that the non-household threshold for eligibility for competition be abolished. It did not recommend that competition should be extended to households. As an interim measure to increase competition, government took action through the Water Supply (Amendment to Threshold Requirement) Regulations 2011 to reduce the threshold in England for eligibility for competition on non-household customers from 50 megalitres to 5 megalitres per year. This increased the number of customers eligible from 2180 to 26,000. The Welsh Government decided to retain the 50 megalitres threshold.

To show the sequence of events, I mention that the review of Ofwat (The Gray Report), which was referred to earlier under Periodic Review, took place in 2011 after the Walker and Cave Reports and before the Government prepared its White Paper.

Substantive proposals were made in a White Paper, Water for Life: Market Reform [46] December 2011. In this paper government decided not to pursue fundamental structural change and would not include legal separation of water companies' retail functions in a water bill, saying that it "did not wish to take risks with a successful model given the challenges" ahead. It would unbundle the combined licence and create a new upstream licence for companies wishing to provide either raw or treated water or to treat or dispose of sewage. Perhaps in recognition of the risks of having less competent companies putting water into distribution systems, it would strengthen the Drinking Water Inspectorate (DWI). It would be a statutory consultee on new entrants.

Following consultation on the White Paper, a new Water Bill was drafted. This Draft Bill related to England alone as the Welsh Government had not yet decided whether to extend competition. At the time of writing the Bill had not yet been placed before Parliament. In connection with competition the Draft Water Bill proposes:

- The unbundling of the combined licence with individual licences for each part of the supply chain;
- All non-household customers, whatever the volume of water consumed being eligible for competing suppliers;
- Allowing retail infrastructures (mains, pipes, storage and treatment) to be connected to the existing primary networks;

- A requirement that Ofwat consults DWI (Drinking Water Inspectorate) before granting a water supply licence. (Presumably this is in connection with concerns over risks that new entrants might introduce incompatible or unsafe water to the networks);
- Access (to networks) prices determined by OFWAT based on costs and a common methodology.

There are other provisions in the Draft Bill, but I focus on the extending competition aspects. Accepting that competition is in principle a good way forward for creating incentives for improved service, what are the specific challenges for making it work in practice. Two questions come to mind, one on whether there is scope for effective competition, and the other related to concern over the maintenance of the monopoly components, the water mains and sewers.

- Will loss of integration and interface costs (contractual, monitoring) of debundling offset the benefits of competition? A report [46] prepared for Ofwat in 2004 stated "the integration of production and water distribution lowers overall supply costs" and "the possible benefits of competition may be offset by cost increases resulting from lost scope economies".
- The underground networks contribute the largest part of the cost of the services. For this reason, and because of congestion under the streets, they have to remain a monopoly. Their dominant cost limits the scope for the benefits of competition, which might be a constraint on progress, but most importantly their maintenance and refurbishment is a critical part of service sustainability. Will Ofwat, with the task of setting access charges, be able to resist the inevitable pressures to reduce access charges below a sustainability level?

It is not expected that the new conditions will be implemented until 2017, so there is a time period for thinking through the details. I believe that it is important to keep in mind that the consequences of loss of a water service are very high on both households and industry. Provision for supplies of last resort must be maintained even it that compromises competition. The trading approach adopted in Queensland, Australia, in 2007 was modified substantially in 2012 (See Chapter 7). I am not aware of any other countries who are introducing such radical changes but hope that the proposals will produce real advances in service quality.

REFERENCES

1. Okonski K. (ed.). (2006). The Water Revolution: Practical Solutions to Water Scarcity. International Policy Press, London. ISBN: 9781905041138.
2. Bakker K. J. (2004). An Uncooperative Commodity: Privatizing Water in England and Wales. Oxford University Press, Oxford.

3. Sjölander Holland A.-C. (2005). The Water Business: Corporation versus People. Zed Books, London. ISBN: 9781842775646.
4. Cook H. F. (1998). The Protection and Conservation of Water Resources: A British Perspective. Wiley. ISBN: 9780471976813.
5. The Development of the Water Industry in England and Wales (2006). Defra and Ofwat 2006 Crown Copyright.
6. Taken for granted. Report of the Working Party on Sewage Disposal. HMSO. ISBN: 9780117502208.
7. Okun D. (1977). Regionalisation of Water Management, A Revolution in England and Wales. Applied Science Publishers, UK.
8. A National Assessment (1977). Report No 4 of the Department of the Environment and National Water Council Standing Committee on Sewers and Water Mains. NWC, UK, June 1977.
9. House of Lords Select Committee on Science and Technology (1982). The Water Industry Volumes I and II. HMSO, UK, December 1982.
10. National Water Council (1982). Water Industry Review. February 1982.
11. Leakage Control Policy and Practice (1980). Standing Technical Committee Report Number 26. Department of the Environment. National Water Council, July 1980.
12. Jordan A. (2000). The Europeanisation of UK Environmental Policy, 1970–2000: A Departmental Perspective. Working Paper, University of East Anglia.
13. Water Supply (Water Quality) Regulations (as amended) (1989). The Stationery Office.
14. Green C. (2000). The lessons from the privatisation of the wastewater and water industry in England and Wales, Berlin water privatisation conference.
15. Water Act 1989. The Stationery Office. ISBN: 9780105415893.
16. Water Industry Act 1991. The Stationery Office. ISBN: 9780105456919.
17. Water Act 2003. The Stationery Office. ISBN: 9780105437031.
18. DWI (Drinking Water Inspectorate) Information Letter 15/2000, www.dwi.gov.uk
19. Water Supply (Water Quality) Regulations 2000 (England).The Stationary Office, ISBN: 9780110188782. The Water Supply (Water Quality) Regulations 2001 (Wales). The Stationery Office, ISBN: 9780110904313.
20. Environment Protection Act 1995. The Stationery Office. ISBN: 9780105425953.
21. Environmental Facts and Figures. Environment Agency, England and Wales. www.environment-agency.gov.uk (accessed July 2013)
22. Directive 2000/60/EC of the European Parliament and of the Council establishing a framework for the Community action in the field of water policy October 2000.
23. The Water Environment (Water Framework Directive) (England and Wales) Regulations 2003 and the Water Environment (Water Framework Directive) (Northumbria River Basin District) Regulations 2003.
24. Introduction to the new EU Water Framework Directive: European Commission. http://ec.europa.eu/environment/water/water-framework/info/intro_en.htm (accessed July 2013)
25. 3rd Implementation Report on the River Basin Implementation Plans. November 2012, European Commission (Environment).
26. Transposition of the Water Framework Directive (2000/60/EC): Defra UK 2003 archive. http://www.parliament.uk/documents/commons-committees/environment-food-rural-affairs/efra-water-framework-directive-2005-updateweb.pdf (accessed July 2013)

27. Check on Progress: Summary of the 2012 WFD classification status for surface waters (PDF, 246KB) Environment Agency, England and Wales 2012. www.environment-agency.gov.uk/research/library/data/97343.aspx (accessed July 2013)
28. Annual Review 2011–12 for England and Wales: Consumer Council for Water 2012. www.official-documents.gov.uk/document/hc1213/hc01/0197/0197.pdf (accessed July 2013)
29. Forward Work Programme for England and Wales 2013–14 to 2015–16: Consumer Council for Water. www.ccwater.org.uk/server.php?show=conWebDoc.2648 (accessed July 2013)
30. Review of Ofwat and Consumer Representation in the Water Sector: Department for Environment, Food and Rural Affairs PB13587. Crown copyright 2011.
31. Service and Delivery: Performance of the Water Companies in England and Wales 2009–10 Report. Ofwat, UK. www.ofwat.gov.uk/regulating/reporting/rpt_los_2009-10supinfo.pdf (accessed July 2013)
32. Levels of Service for the Water Industry in England and Wales 2004–5 Ofwat September 2005. www.ofwat.gov.uk (accessed July 2013)
33. Setting Water and Sewerage Price Limits for 2005–10: Framework and Approach. Ofwat, 2003, www.ofwat.gov.uk (accessed July 2013)
34. Reporters to Ofwat: Reporters Protocol. Ofwat 2003. www.ofwat.gov.uk (accessed July 2013)
35. Water prices in England and Wales from April 2007 (2007). Water UK Press, Release 26 February 2007, www.water.org.uk (accessed July 2013)
36. Water Management. House of Lords Science and Technology Committee Report (2006). The Stationery Office. June 2006. ISBN: 9780104008713.
37. Singh U. (2006). A Personal Communication.
38. Response of the Water Services Regulation Authority (Ofwat) to the House of Lords Science and Technology Committee Report on Water Management, www.ofwat.gov.uk
39. The Drought Direction 2011 archive. http://archive.defra.gov.uk/environment/quality/water/resources/documents/droughtdirection2011.pdf (accessed July 2013)
40. Drought Permits and Drought Orders: Information from the Department of Environment, Food and Rural Affairs, Welsh Assembly Government and the Environment Agency May 2011 archive. http://www.gov.uk/government/uploads/system/uploads/attachment_data/file/69441/drought-permits-drought-orders.pdf (accessed July 2013)
41. Assessment of regulatory barriers and constraints to effective interconnectivity of water supplies: R&D Technical Report WT0921/TR Defra, September 2010. http://archive.defra.gov.uk/environment/quality/water/resources/documents/wt0921-technical-report.pdf (accessed July 2013)
42. The Independent Review of Charging for Household Water and Sewerage Services: Published by the Department for Environment, Food and Rural Affairs. PB13336. http://archive.defra.gov.uk/environment/quality/water/industry/walkerreview/documents/final-report.pdf (accessed July 2013).
43. Independent Review of Competition and Innovation in Water Markets: Defra. ISBN: 9781845325688, https://www.gov.uk/government/uploads/system/uploads/attachment_data/file/69462/cave-review-final-report.pdf (accessed July 2013)
44. Tackling Bad Debt in the Water Industry January 2012: www.gov.uk/government/consultations/tackling-bad-debt-in-the-water-industry (accessed July 2013)

45. Draft Water Bill July 2012, ISBN: 978-0-10-183752-1. www.official-documents.gov.uk/document/cm83/8375/8375.pdf (accessed July 2013)
46. Investigation into evidence for economies of scale in the water and sewerage industry in England and Wales. Report commissioned by Ofwat from Stone & Webster Consultants January 2004, http://www.ofwat.gov.uk/pricereview/pr04/rpt_com_econofscale.pdf (accessed July 2013)

Chapter 10

Summary

Bringing it all together

Each of the previous nine chapters has dealt with an important area, giving my views on the key elements for sustainable water services. Each area needs to be addressed in the provision of sustainable water services. I hesitate to label any of my views as representing recommendations, since it is for the reader to consider the strength of the arguments based on the information given. This chapter provides easy access to the views expressed in each chapter under the chapter titles. It is included largely to provide students with a checklist for revision, but other readers might wish to use it as a means of obtaining quickly an overview of the issues. I make no apology for the repetition of some views in a number of contexts. The following statements are given as short bullet points, without any of the supporting information given in the chapters in which they are discussed. Readers should refer to the relevant chapters for more information.

CHAPTER 1 PART 1: IMPORTANT POLICY ISSUES

- There should be an integrated approach to policy and planning of water and sanitation services, all within one government department, or with a designated lead department having authority to lead and coordinate.
- Policy, regulation and delivery should be separated to give focus to the application of the required expertise, and to provide transparency.
- Water service operations should be managed on a large enough scale to attract high quality management, and to afford competent scientific support.
- Decentralisation is not about scale of operations, but rather allowing local management to operate without political interference. However, there has to be local capability to manage, and decentralisation steps must be accompanied by capacity building and training.

- There can be a policy conflict between optimisation of integrated water resource management (IWRM) and the shorter-term priority to extend access to water and sanitation services. For sustainability, optimisation of IWRM is essential, but generally should not be the main driver.
- The delivery of drinking water should be an integrated process. As a minimum, water intakes, water treatment and water distribution should be under the same management. This argument is now being tested by attempts to introduce direct competition.
- Water supplies can and should be continuous (24/7), with pressure maintained in distribution systems. Rationing by intermittent supplies has been shown to be ineffective, require more, not less, source water, and pose a risk to health. In recognition of this 24/7 is increasingly being adopted as policy.
- Policy on the provision of sanitation is not about a choice of technical ideologies, but about having an understanding of the objectives of a particular service, and designing a system to meet those objectives.
- Corruption is a symptom of some established social systems which need reform. It is associated with policies which result in under-recovery of costs of water services, with associated poor pay for management and staff.
- Full-cost recovery is essential for the sustainability of water services, without which the infrastructure falls into decay.
- Safe drinking water is necessary for survival, but it is also important that people drink enough water to remain fully hydrated to achieve their maximum mental and physical potential.

CHAPTER 1 PART 2: WATER AND SANITATION AS A HUMAN RIGHT

- The UN Resolution of Water and Sanitation as a Human Right should not be seen as an end in itself, but as an opportunity to make progress on the road to universal access.
- A wholly legally-driven approach to implementation would result in chaos: handbooks to provide guidance on practical implementation are in preparation.
- I have given my views on some key issues. Please see the chapter for my comments.
- At the time of writing, work is in progress on the development of post-2015 goals. They are likely to be influenced by human rights considerations.
- Implementation of the Resolution will be highly financially and logistically demanding. It will take time, requiring long-term planning based on progressive realisation. I make the plea that dogma and narrow interests don't get in the way of progress.

CHAPTER 2: REGULATION AND REGULATORY BODIES

- Regulation is the process of interpreting and implementing laws, policies and regulations, to achieve the original objectives.
- One definition of economic regulation is "the rules and organisations that set, enforce and change the allowed tariffs and service standards for service providers".
- Generally, self-regulation has failed to produce efficient and sustainable water services.
- An independent body, with responsibilities for economic regulation, is necessary to ensure a sense of realism in the setting of objectives, and to oversee the process of turning objectives into deliverables, whether the service providers are public or private.
- There are various models used for tariff setting. Incentive-based approaches are tending to replace rate of return regulation.
- An independent drinking water quality regulator can ensure that there is transparency in the reporting of drinking water quality, and in the investigation of incidents affecting drinking water.
- An independent environment regulator avoids the risk of decisions being made "behind closed doors" for political expediency.
- A well-defined contract provides the means of controlling the delivery of the desired performance of a water service operator, whether public or private, but it falls short of effective regulation. Effective regulation requires transparency, public participation and scrutiny of the contract owner.
- Regulatory bodies can take many forms. Ideally, regulators should be single function. However, there are good examples of multi-utility regulators, and of regulators combining economic and quality regulation. Clarity on responsibilities is essential.
- Economic regulators should have the authority to set tariffs, based on the need for sustainable cost recovery, and be free from political interference.
- Public operating utilities should not have the role of regulating others. This creates conflicts of interest, particularly related to performance comparison. Also, regulation requires different skills.
- Regulators can benefit from external inputs to develop and maintain expertise, either from expert panels or through regulators' networks. Expert Panels can enhance credibility, providing the members are independent and not chosen "politically".
- In those developing countries in which governments do not have a policy of establishing formal independent regulators, local water committees, supported by NGOs, can provide a form of regulatory system. Such approaches should be developed in line with the UNDP Human Development Report 2006 which, although largely forgotten, provides sound recommendations.

- In meeting the needs of the poor, there has to be recognition that their main difficulty is not the ability or the willingness to pay, but the inability to save. For both one-off connection charges, and payment of water charges, systems of pay-as-you-go are needed.
- General subsidies result in unsustainable systems, and benefit the "rich" more than the "poor". Subsidies to assist the poor should be directed at one-off costs, such as extensions to distribution systems and connection charges.

CHAPTER 3: PLANNING, FINANCING AND COST RECOVERY

- Water is a human right, but that does not mean that water services can be free.
- Insufficient cost recovery leads to a downward spiral with deteriorating water services (see Figure 3.1).
- Costs of water services can be considered as start-up costs, operating costs and the cost of renewal of the infrastructure.
- Paying for start-up costs, including house connections is difficult, especially in developing countries. There is a need for one-off subsidies, or a means of spreading the costs over a long period of time, say 25 years.
- Where there is already high access coverage, the costs of connecting the remaining population, which is often the poor sector, can be recovered from the general water charges levied on all consumers.
- Donor money, or bank lending, to assist in paying for start-up costs are important, but equally important is assistance in the refurbishment of old systems. Donors should give greater emphasis and priority to the refurbishment and extension of water distribution systems.
- Asset management planning should be used to predict future investment needs.
- Water charges should be sufficient to recover operating costs and to make provision for the maintenance and renewal of the existing infrastructure.
- General subsidies, which give greater benefit to the rich than to the poor, are generally unsustainable and result in a lack of maintenance of the infrastructure. Subsidies should be directed to the poor through the most focussed mechanisms possible.
- The poor are paying more than the "rich" for non-tap water supplies. Providing their connection charge is subsidised, they will end up paying less than before, even with cost-recovery tariffs. So the poor can pay but they are unable to save money. Therefore, they need "pay as you go" systems.
- Experience around the world has shown that the threat to cut-off supplies to non-payers is essential for sustainability. Examples show that generally non-payment is not a matter of being unable to pay, but rather of being unwilling to pay. However, there is a need for some means of providing water to the genuine cases of inability to pay.

- Tariffs should be set by independent regulators. Politicians are rarely willing to set tariffs at the level necessary for sustainable water services.
- Integrated planning and tariff setting is necessary to avoid an imbalance between water resources and water supply, and between the level of service and affordability.
- An integrated planning/tariff-setting process ensures that the costs of political objectives are included fully in the tariff-setting process. An integrated process identifies unrealistic ambitions and objectives. Planning processes should allow for changes to delivery timetables to be made, to bring investment programmes into affordable timescales.
- With rapid urbanisation in developing countries, water and wastewater planning should be an integral part of city planning. More research is needed on development of water distribution systems as part of cities of the future.

CHAPTER 4: DRINKING WATER QUALITY STANDARDS AND REGULATION

- Access to water should mean access to safe water with consideration of drinking water quality.
- Standards cover microbiological, chemical and radiological risks. The risks from pathogens are by far the most important.
- Numerical standards are important, but the emphasis should be on approaches to managing risk.
- The Bonn Charter, with the goal of "good safe drinking water which has the trust of consumers" is a set of high-level principles for the supply of safe drinking water. The WHO Guidelines for Drinking Water Quality provide both the framework, and the most rigorously prepared guidance for drinking water quality standards.
- Together, the Charter and WHO Guidelines for Drinking-Water Quality, provide the basis for government policies and for the establishment of drinking water quality standards. Both advocate the use of (drinking) water safety plans.
- Water safety plans are a means of assessing risks to drinking water quality and the management of those risks. They can be applied equally in developed and developing countries, although the approaches and level of detail need to match the local circumstances.
- Although setting health-based targets is desirable, it is difficult and time consuming. In developing countries, I do not see health-based targets as necessary to justify investment in basic water and sanitation services and hygiene. It is known already that the benefits greatly exceed the costs.
- Water safety plans should cover the management of risks from source to mouth. This should include steps to manage risks at households.
- Water safety planning is increasingly a regulatory requirement.

- Standards are a combination of treatment standards and numerical standards for individual parameters.
- Dealing with the risks from pathogens is paramount, and disinfection should not be compromised due to concern over the low risks from by-products of disinfection.
- In developing countries, where it may take many years and high investment to achieve standards based on WHO Guidelines, it is sensible to set time-related achievable interim standards. These provide the opportunity for a sense of achievement and avoid a sense of hopelessness. The first level interim standards can be based on protection from pathogens alone unless there are concerns over chemicals at toxic concentrations.
- Magnesium in drinking water is now recognised as an important factor in reducing heart disease and strokes. This has implications for the specification of desalinated and re-use water, and in the use of those householder filters which remove magnesium.
- There is little point in drinking water standards unless they are monitored. There is little point in monitoring unless there is enforcement. Enforcement cannot be effective unless there is a means of paying for improvement programmes. This means that drinking water quality regulation must be part of an integrated planning and tariff setting process.

CHAPTER 5: TRANSPARENCY AND PUBLIC PARTICIPATION

- Transparency is important, not only as a means of combating corruption, but also to prevent the concealment of politically unpalatable information; for example water revenue being used for other purposes, or water quality incidents being hidden from the public.
- Transparency and public participation go hand in hand, with strong public participation strengthening the position of authorities, rather than diminishing their power.
- Public participation provides the opportunity for public understanding, and support for investment which has an associated impact on water service charges.
- Public participation can be at a number of levels from the least involved level of information provided with water bills, to the most involved level of communities owning and managing their water services.
- Public participation is increasingly seen as part of achieving accountability of service providers. The traditional route, the so-called long route is via politicians, whereas a faster response can be obtained through direct communication between citizen and service provider. A number of tools are used. Citizen report cards have been shown to be very effective

- In the developed world, with consistently good water services, the public may be satisfied with the knowledge that there is full and good information available. They are likely to be satisfied that there are independent regulators and consumer associations, who will look after their interests and report any problems. The situation depends critically on whether the media are confident that there is full transparency, as any suspicion that something is being hidden will undermine the credibility of the authorities.
- However, in the developing world, public participation may be necessary to provide the stimulus for better services, and the means by which improvements are achieved.
- For public meetings to be effective, they need to be held on a regular basis to achieve trust and transparency. Meetings held only when there is something controversial to discuss, such as a new reservoir, are liable to become confrontational.
- Water consumer councils, operating either as part of an independent regulator, or preferably wholly independent, provide an effective voice for the public. They are able to make informed consumer inputs on water tariffs and willingness to pay, and they can seek compensation for consumers for poor service.
- In the developing world, there are many good examples of more direct public participation resulting in increased access to water services, both water and sanitation. Although history has shown that trying to pursue direct replicas of these examples does not necessarily work, there are some definable ingredients for success. They are:
  - government support for local initiatives;
  - strong and effective community leadership;
  - a respected and knowledgeable NGO;
  - appropriate choice of technology;
  - full-cost recovery.
- The most important ingredient may be a means of achieving government "accountability to its people". This is the objective of the WaterAid-led Citizens' Action initiative.
- Social media is changing the way in which organisations and citizens communicate. Official information increasingly has to be timely, accurate and available through social networks.

CHAPTER 6: STIMULATING IMPROVED PERFORMANCE

- Without separation of policy and delivery functions within local government, it is difficult and probably impossible to achieve highly efficient operation of water services.

- Incentives for management to perform better in water sector operations should be tied to output performance of the utilities, with publication of performance objectives and performance figures.
- It is important to note that the term benchmarking was used originally to describe the method of learning from the "best in class". It is now also used to describe systems of measuring relative performance of organisations. To differentiate between the two functions, the latter is now called metric benchmarking, and the former is called process benchmarking. Metric benchmarking uses a relatively small number of measures compared with process benchmarking, which is much more detailed.
- Effective metric benchmarking requires careful selection of the measures to be used to direct managements' energies towards achieving the desired utility performance. The starting point should be clarity on priority of objectives. It should not be a list of measures used elsewhere. There should be a few key measures and a focus on obtaining reliable and meaningful data.
- Initially, effort should be focussed on obtaining good baseline data. Without meaningful data, it is not possible to measure progress.
- Published "league" tables of performance are a useful incentive to management, but it is important that results are qualified to take into account the differences in history and operational conditions of each utility.
- Trends in performance of each utility, which show improvements over time, can be more useful than absolute performance figures.
- The International Benchmarking Network for Water and Sanitation (IBNET) provides advice on setting up benchmarking schemes and promotes international comparisons.
- There are a number of good examples of benchmarking schemes in both developed and developing countries, most of which are based on metric benchmarking. But league tables can be threatening, with the numbers receiving greater focus than improvement of the service. One successful approach is to start with process benchmarking, with utilities encouraged to learn from best of class. In time there will be the confidence and trust to move successfully to metric benchmarking.
- A public service ethic is vital in the protection of public health, but in the past there has been a tendency for public water service managers to have a job for life, often resulting in complacency. Continuation of tenure should be subject to performance, in a similar way to that required through a private-sector contract.

CHAPTER 7: STEPS FOR SUCCESSFUL PUBLIC SECTOR OPERATIONS

- The traditional municipal model does not meet important criteria necessary for effective and efficient water service operations. There are some good examples of cities successfully achieving reforms within the municipal model, but generally it is necessary to change to a different governance structure.

- Although much can be achieved through bottom-up actions, as is seen in Chapter 5, national governments should take the lead through legislation, or through providing incentives for reform. A key requirement is to separate water service operations from local government policy functions.
- Separation of policy and operations can be achieved in a number of ways. It can be accomplished through internal contracts as in the cases of Seattle and Brisbane. However, most commonly, it is achieved through cities forming so-called corporatised-public companies. These companies are owned by the cities, but are required to perform as independent companies operating under private-sector rules. Examples are The Netherlands, parts of Australia, and Scotland in the UK.
- Other means of achieving separation are through the letting of concession contracts, as is common in France, or through the formation of equity-based private companies, as in England. See Chapters 8 and 9.
- Reforms can come about following a major incident, as in Ontario in Canada. This case resulted in greater government control rather than reform of the sector, and will provide an interesting comparison with other solutions in the future.
- The development of public service operations can include the introduction of independent regulation, equivalent to that used to regulate private sector operations. I argue that regulation, with independent tariff setting, is just as important in a public sector structure as in privatisation. Examples are Scotland, and the State of Victoria in Australia. Another example is Ghana, although in that country the regulator can only advise on tariffs.
- Success in Singapore required continuity of government pursuing well-defined policies implemented through long-term planning. It is important to recognise that, at the beginning of the reform period, Singapore, when a relatively poor country, had problems of intermittent supplies and a highly polluted water environment faced by many countries today.
- The examples of Phnom Penh and Uganda show the importance of having good managers allowed to operate without political interference, but within a climate of political commitment and sound governance. These successes should give confidence that it is possible to establish sustainable water services for all.
- It is important to recognise that reform takes time, as can be seen in the cases of Phnom Penh, Uganda and Haiphong in Vietnam. It is difficult to recover from a situation arising from a downward spiral of decline. A poor service both arises from and results in low cost recovery. Political interference in management results in low motivation and lack of management initiatives. A deteriorated infrastructure requires a high level of investment over a long

period of time. All of these political/management issues have to be tackled in an integrated way. In developing countries, outside assistance can be most valuable, especially in the early period through financial and technical support. The Haiphong example shows what a perceptive and knowledgeable donor can achieve, if able to provide support over the whole reform period.

- The examples of public sector reform described in Chapter 7 show that success is dependent on putting some important elements in place. The most important is sustainable cost recovery. This is not achievable without independent tariff setting, as experience shows that politicians are unwilling to raise tariffs to the required levels. Although, as in The Netherlands, sustainable cost recovery can be achieved through local authority action, generally there is a need for an independent regulator.

CHAPTER 8: STEPS FOR THE SUCCESSFUL USE OF THE PRIVATE SECTOR

- The private sector is almost always deeply involved in water services, in design, construction, and in the provision of other support facilities for utilities. Private sector firms have to make a profit to stay in business, and generally they provide services more efficiently than if they were done in-house.
- There are emotive arguments against the use of the private sector for management and operations, with opponents seizing on controversial elements, such as necessary tariff increases. The emotive arguments tend to be associated with the view that water is a human right, and therefore should not be for profit.
- A more compelling argument is that since water supply is a monopoly due to the high cost of the distribution infrastructure, it is not possible to have effective competition with the market controlling water pricing. The UK is proposing to provide all non-household customers with a choice of supplier but, generally, it can be expected that water services will remain a monopoly.
- There are alternative emotive arguments, with the view that the public sector is inherently less efficient than the private sector, as there are fewer incentives to strive to do better.
- Competitive tendering for management or concession contracts creates periodic competition, but the choice is with the contract owner, not the consumers. Contract periods should be long enough to give some management continuity and sufficient time for recovery of charges, but the longer the period between rebidding the less opportunity there is for competition. It is suggested that generally contracts should not be longer than 15 years unless there is high capital investment involved.

- A study of controversial "concession" contracts shows that the "failures" can be attributed to the way in which the contracts were set up and managed, with equal blame falling on the authorities and the private companies. It can be argued that all the blame for failures should fall on the authorities letting the contracts, as they are responsible for delivery of the service, whether managed in-house or contracted out. This is the argument for regulating the contract owner, and not the contractor.
- One of the most common complaints is that water service charges are increased so that the private sector can make "big profits". What tends to happen is that water charges have been too low for sustainable systems, so there has been insufficient investment in systems, and authorities need to increase charges to pay for the improvements. This would be necessary whether or not the private sector was involved, but the blame falls on the private companies.
- Effective use of the private sector requires a number of important elements:
  - clarity on policy, particularly in relation to water service charges and the use of subsidies to assist the poor;
  - public participation, particularly in developing countries, in assisting in the development of policy on aspects such as access charges, and methods of payment;
  - transparency, so that the public understand how the contract aims to improve the service, what will happen to charges and subsidies, and how the contractor's performance will be monitored. This is best achieved through some form of independent regulation or supervision;
  - an effective procurement process, based on good information, and with provision for periodic adjustment of contracts to cater for changing circumstances. In particular, any potential problems, especially those associated with poor information, should not be "fudged" but highlighted with the necessary collection of information included in the contract. See Chapter 8 for more information on procurement.
- A major consideration is the allocation of risk, based on the important principle that each risk should be taken by the party most able to manage it. For example the contractor should take risks associated with construction and operations, and the authority should take risks associated with changes in government and legislation. These allocations should be covered explicitly in the contracts.
- Partnering, in which two parties enter into an agreement to deliver something, working together to solve problems within a climate of trust, has been very effective in Scotland. It is a true public-private partnership.
- A perfect concession contract may not be achievable but the inclusion of some key ingredients, given in Chapter 8, will avoid the problems which have arisen in some contracts which failed.

- There is a need for some independent scrutiny of operations under concession contract conditions. Such regulation should not interfere with the detail of the contracts, but provide objectivity and transparency on which contracts are based.

Equity-based privatisation

- To be clear about the difference between private contract privatisation which has been covered above and equity-based privatisation, equity-based privatisation involves the sale of assets to private companies, which are licensed to provide water services for specified periods of time. Such licences include the requirement to maintain the serviceability of existing assets and to finance, build and operate any new assets required to meet defined objectives.
- Chapter 8 gives the very successful example of Chile, in which there is a combination of equity-based private sector involvement and concession contracts. Chile has in place all the important elements for success: strong regulation, sustainable cost-recovery, and targeted subsidies for the poor.
- Chapter 8 also gives three examples from China, a successful concession, a poorly designed BOT contract, and a successful equity and management based joint venture. The third example is seen in China as a model for other cities to emulate.

CHAPTER 9: THE STORY OF ENGLAND AND WALES

Chapter 9 gives an account of the history of water services in England and Wales from 1945 to the present day, together with a selection of references. This summary gives some of the key points and the lessons to be learnt from the experience of the developments in England and Wales.

- History is important in the success of equity-based privatisation in England and Wales. There was some consolidation of water supply operations following the Water Act 1945, reducing the number of utilities from over 1000 at the time of the Act, to 198 in 1973. In 1974, 1620 water service units, consisting of water supply operators, sewerage operators, and river authorities, were consolidated into 10 river-basin based regional water authorities (RWAs), giving the required scale of operations and integration of functions to provide the opportunity for efficient operations. At the same time, national grants ceased, leading to the requirement for full-cost recovery through charges.
- The change in 1974 to river-basin structured water authorities was heralded as the "ideal" way to manage rivers and water services. However, due to the inheritance of a lot of debt, and Government restrictions on borrowing, the RWAs probably did not meet the expectations of those who championed it.

- There were many aspects requiring attention, including investment in drinking water quality, and the need for greatly improved approaches to dealing with wastewater, both in discharges to rivers and the sea. European Directives were introduced requiring the introduction of significantly improved standards. Against this background, there was a downward trend in investment during the period 1974–1982. In 1977, the Government introduced equalisation of charges, to provide for more equal charging across the country. This discouraged any RWA from striving to be efficient, and the law was repealed in a change of Government in 1979.
- In 1982, the RWAs became more business-orientated, with a change from Boards made up of local authority representatives to Boards constructed on a private-sector model. This important stage, which can be considered as the corporatised-public stage, provided a business culture able to adapt to the demands of privatisation.
- The Water Act 1989 provided for the flotation of the 10 RWAs on the stock market, with much of the stock purchased by small investors. At the same time, the 26 Statutory Water Companies, some of which had been in existence for over a hundred years, were brought under the same regulatory regime.
- Privatisation brought independent economic regulation, with the formation of Ofwat, but with policy remaining with Government. Ofwat is able to fix tariffs objectively, based on financing requirements, full-cost recovery, and efficiency targets, free from political interference.
- One of the flaws in the RWA structure had been that water service operators were responsible for regulation of river quality, which was resented by some of the other industrial dischargers. On privatisation, responsibility for river regulation was given to a new body, the National Rivers Authority, which later became part of the Environment Agency.
- Although the "purists" felt that privatisation destroyed the ideal RWA structure, river basin management continues under the Environment Agency, which is an independent regulator with responsibility for both water resources and the environment.
- Drinking water quality regulation became the responsibility of the Drinking Water Inspectorate, an independent regulator, but administratively part of the Department of Environment, Food and Rural Affairs (DEFRA).
- Privatisation, by removing the constraints and limitations of Government financing, provided the conditions under which sufficient investment money could be raised from the private capital market. This provided for the requirements of European Directives to be met, and for a start to be made on refurbishing the ageing infrastructure.
- The regulatory system has worked very effectively, with a combination of the integrated planning periodic review process, and comparative competition, delivering both improved services and increased efficiency.

The main lessons from the England and Wales story (so far) are:

- Consolidation provides viable operating units able to attract high-quality management and to afford good scientific support.
- Even a "perfect" structure, as some saw in the formation of the RWAs, will not deliver the required outputs, if it is constrained by inherited debt and inappropriate government policies.
- A move to full cost recovery from charges is important, and should take place well ahead of either corporatisation or privatisation.
- An initial move from a municipal-type organisation to a corporatized public company structure provides a good basis for the preparation of utilities for successful privatisation.
- Comparative competition is an effective substitute for direct competition in a monopoly situation.
- Strong independent economic and quality regulation, working through a periodic integrated-planning process, delivers high performance.

But the development of the story will continue. There were concerns over security of supply under drought conditions; consumers in the South West area, due to a low density of population and large investment requirements had much higher water bills than elsewhere in England and Wales; debt levels increased with increasing numbers of household consumers not paying (perhaps largely choosing not to pay) without fear of cut-off; and there were pressures from large users to have a choice of supplier. Government responded by commissioning three major studies. A discussion of these reports and subsequent government actions are given in Chapter 9. At the time of writing bills are awaiting parliament. Assuming that they become law, they will take England into a new era of competition with some question marks on the outcome.

FOOTNOTE ON THE PUBLIC VERSUS PRIVATE QUESTION

As can be seen from Chapters 7, 8 and 9, all forms of water service operation can be successful, providing that there are sound government policies, some form of independent regulation or scrutiny, (or excellent management with political support) and incentives to perform better. The public-private argument is unproductive, and detracts from those policy matters that are important in the achievement of effective, efficient and sustainable water services.

Appendix

The UNDP human development report 2006

Taking forward the Report's recommendations on governance and regulation

Second Edition Comments

This was an important document which was receiving a lot of attention at the time I was writing the first edition of this book. Although it appears that it is no longer receiving attention as a driver for progress, its messages are as valid, at the timing of writing this second edition (2013), as they were in 2006. The attention given in the report to governance and regulation continues to reinforce the content of this book, so it has been retained as an appendix. The first edition text remains unchanged.

First Edition Text

I have mentioned this report a number of times in the book. It is important that some key recommendations in the report are taken forward. I refer below to a number of statements in the report which relate to governance and regulation:

- The establishment of dedicated water and sanitation ministries, led by senior cabinet ministers, would create a political structure capable of overcoming the fragmentation of policy and the resultant underresourcing.
- Each country should have a national water and sanitation plan.
- The need for the development of guidance on establishing effective regulation.
- Increase in capacity building in policy and regulation at national and local government levels.
- All water providers should be bound by equity performance targets, stipulating goals for extending access to poor households. The targets should include clear provision to unserved urban and rural communities, the expansion of standpipe provision in slums and the delivery of free or low-cost water to low-income households. Contracts drawn up within

public-private partnerships should include targets in these areas, with full public disclosure, monitoring by an independent regulatory body and penalties for non-performance.

- A pro-poor expansion package should include lifeline tariffs, cross-subsidies that transfer resources from higher-income to low-income households through utility pricing or targeted fiscal transfers, and sustainable and equitable cost-recovery measures.
- Service providers should set charges to cover recurrent costs, with public finance covering capital costs for network expansion.
- In water and in sanitation, as in most areas, governments work best when they work in partnerships that build on the energy, drive and innovation at a community level – and when they listen to people.
- Water and sanitation service delivery brings together a wide range of providers and extends across complex markets. Governments have a responsibility to ensure that providers and markets are governed to prevent the abuse of monopolistic power and to deliver safe, affordable and reliable water and sanitation to the poor. One of the problems with current regulatory frameworks is that their remit does not extend beyond large-scale formal providers.
- This is a broad agenda. It goes beyond the narrow preoccupation with private and public ownership that has dominated debates on water and sanitation. While these debates have highlighted important concerns, they have diverted attention from important public policy issues.
- The financing, delivery and regulation of water and sanitation services pose tough public policy challenges that cannot be resolved simply by claiming that water is a human right, or by debating about public or private operators.

It is refreshing to see the above statements in the report, asserting that the public-private argument has diverted attention away from the important issues. It is also good that there is recognition of the need for full-cost recovery of "recurrent" costs through charges, with public finance used for network expansion. The need for regulation features strongly, but how can this be delivered? I am not pretending that it is easy. There is no substitute for putting in place some fundamentals for sustainable water services. All aspects have been covered in the preceding chapters, and I have attempted to summarise them below.

It is desirable to have a dedicated Ministry covering all aspects of water and sanitation policy. At the very least, there should be a designated lead Ministry responsible for managing and coordinating policy development. Policy development should include how to establish operations on a scale large enough to attract good management. There should be separation of service delivery from government so that managers can operate without political interference. Governments should address the subject of cost recovery through removing general subsidies, and should direct public finance to extend access to the unserved. This should include payment of access costs. There should be recognition that, as it is difficult for the poor to save money, they need a method to "pay as you go". As they are paying much more for their water than those with higher incomes who are connected to a network, there may not be a need to subsidise the payment of charges. However there should be a safety net for those who are unable to earn money – this can be done, as described in examples given in Chapter 5, through free standpipe supplies.

Independent regulators can and should be established to regulate both water tariffs and where relevant, sewage tariffs and water quality in urban areas. In small countries, these can operate at a national level and, in larger countries, at a regional level. These regulators should be given the authority to manage the planning process, set tariffs and enforce standards. The main difficulty here is governments' unwillingness to relinquish the authority, not appreciating that the existence of an independent regulator allows them a much stronger position with the public. Government's role should be policy and setting objectives. Government should decide on priorities for improvement and Government should decide on subsidies for the poor. However, there is a need for an independent body to make an unbiased assessment of whether the objectives are realistic and deliverable in the proposed timetable, and with the associated cost and tariff implications. Independent regulators tend to have the trust of the public, and are able to explain why something cannot be achieved more quickly, and why water tariffs have to be increased to achieve sustainable water services. There is no reason why these regulators should not have a wider remit. They can cover tanker services as in Ghana. They can also cover rural areas, providing that an over-complicated approach is avoided. I will return to the needs of rural areas and suggest an alternative to a formal regulator.

As mentioned, a regulator needs to be independent and to have close links with the public, so what is the "specification" for a regulator in developing countries. There is no single specification, as each system needs to be tailored to the needs of each country, but it should include the following characteristics:

- There should be a Regulatory Board, chaired by someone with status in the country, but who is not a politician. If there are no recognised public groups with which the Regulator can work, the Board might include representatives from consumer or community groups.
- It should have responsibility for the planning process on which improvement programmes are based and which determine tariffs. This responsibility should include managing the process of developing and maintaining national water and sanitation plans.
- It should have responsibility for setting tariffs and monitoring the delivery of improvement programmes. Systems in which the regulator has an advisory role, and politicians take tariff decisions, do not work very well.

- It need not be expensive. There is no need for the regulator to employ the number of staff required to operate a regulatory system in the developed world. The work of NWASCO in Zambia demonstrates what can be achieved with political will and dedicated management and staff. It is sufficient to have a small number of "resident" staff, who manage the regulatory processes, with specific expertise bought in as necessary. The aim should be to keep it simple to minimise information requirements. Tariff calculations, which are 90% accurate using basic cost data, are likely to be more valuable than an attempt at complicated modelling which is not supported by available data.
- The regulator should see its role largely as facilitating planning, tariff setting and communications, rather than acting as a policeman. However, if the service providers choose not to co-operate, the regulator should have the necessary powers to take enforcement action.
- In its establishment, it should seek input from those with experience in regulation to help create an appropriate regulatory culture, but avoid copying another country's system.
- The regulator should be part of a regulators' network in which experience can be shared. Ideally, it should be a part of a network of regulators operating under similar conditions.

Returning to the question of regulation in rural areas, if it is not feasible to use an "urban" regulator, there are good examples of community groups taking on a combined overseeing and tariff fixing role, supported by knowledgeable NGOs. In Chapter 5 I gave the example of Kolhapur in India, in which four villages took on the management of their supplies and achieved virtually 100% revenue collection. In some respects, it is easier in rural areas because they are more cohesive communities. Communication in villages is easier, so people are able to learn why sustainable cost recovery is necessary, and peer pressure minimises non-payment.

All of the above requires governments to give water and sanitation priority and to enable the necessary reforms to take place, but how can unresponsive governments be persuaded to take action? In Chapter 5 I refer to the Water Aid initiative called Citizens' Action. This helps communities to engage with service providers and government, with local NGOs facilitating the engagement process. This is proving to be successful in stimulating policy action.

In the Report, it is recommended that a global organisation, similar to the Global Fund to fight AIDS, Tuberculosis and Malaria, is established. I am sceptical about the need for yet another international body, but I do see value in donors getting together and agreeing on the importance of persuading national governments of the need for independent regulation. Support should be given to capacity building of policy makers and regulators. Working with ministries in recipient countries, this should include capacity building in the preparation of national water and sanitation plans, in which key policy areas such as the establishment of independent regulators and cost recovery would be addressed. Support to regulators would aim to provide them with a regulatory framework, built on cooperation with service providers, in the implementation of policies. A regulatory body needs to be tailored to the needs of its country, but donors could agree on and promote a model regulator framework for developing countries. NWASCO in Zambia would be a good starting point.

Index

K

L

M

N

IWA Publishing's authorised EU representative for General Product Safety Regulations is Diane D'Arras, 15 rue Duret, 75116 Paris, France, e-mail: safety@iwap.co.uk

Printed and bound by CPI Group (UK) Ltd, Croydon, CR0 4YY
16/06/2026
02140365-0002